L. L. Cavalli-Sforza

ELEMENTS OF
HUMAN GENETICS

SECOND EDITION

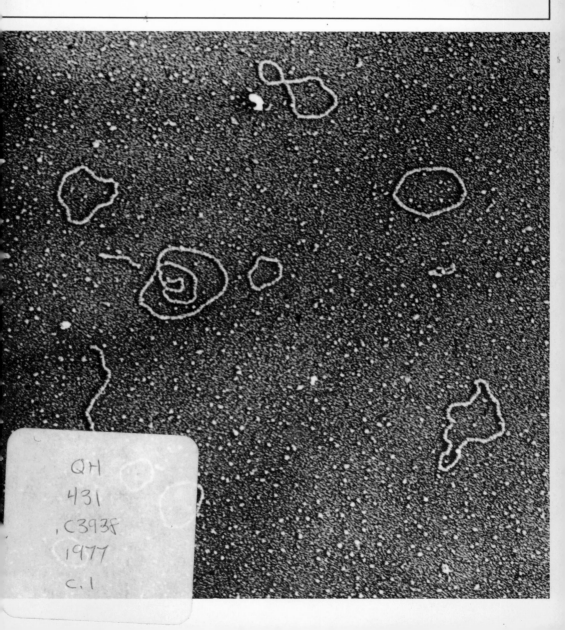

SECOND EDITION

ELEMENTS OF
HUMAN GENETICS

About the Cover:

Research in genetic engineering. Restriction enzymes open DNA at any of a number of specific sites. These tend to join, forming circles. The figure is obtained by electronic microscopy of virus SPP1. Depending on the site at which breaks in DNA have occurred, shorter or longer fragments have formed, from which smaller or larger circles of DNA have originated. (Courtesy V. Sgaramella.)

SECOND EDITION

ELEMENTS OF HUMAN GENETICS

L. L. Cavalli-Sforza
Stanford University School of Medicine

THE BENJAMIN/CUMMINGS
PUBLISHING COMPANY

Menlo Park, California • Reading, Massachusetts
London • Amsterdam • Don Mills, Ontario • Sydney

ISBN 0-8053-1874-7
   FGHIJKL-AL-79

Contents

Preface

Interest in human genetics has increased of late and is bound to continue increasing. It is not exaggerated to say that we are all potentially interested in human genetics, as reproducing individuals and as individuals who are concerned about health and about causes of individual and group differences. Until recently, the study of genetics took place mostly on animals and plants, or lower organisms such as molds or bacteria and their viruses. Most of the conclusions on genetic mechanisms that are thus obtained, fortunately, carry over from one organism to another, including humans. But today much of this knowledge can be obtained directly from humans, and this can only increase the relevance of genetic studies to human affairs.

There is thus a potentially wide audience for the study of human genetics, which is somewhat restricted, in practice, by the difficulties inherent to the subject matter. Genetics uses more abstract thinking than most other biologic disciplines. It thus puts a serious challenge to the writing of a book such as this, which is directed to a wide audience. I have made a deliberate effort to simplify the presentation to a level that does not require knowledge of biology or chemistry. It is assumed, however, that words such as "molecule" or "cell" can be considered as part of general knowledge; I have made only a perfunctory effort at defining a cell, as readers will easily perceive by reading the beginning of Chapter One, and no effort at all at explaining a molecule. Also, the use of numbers is so moderate that most "number-blind" readers should not be offended. The experience gained using this book in its first edition, as an Addison-Wesley module, and in foreign translation indicated that the major aim of being understandable by a wide audience of college freshmen, as well as high school teachers and professionals, was achieved with no serious loss of rigor.

Chromosomes are explained in Chapter One; the description of mitosis and meiosis is kept to the absolute minimum necessary for understanding Mendelian genetics. Information on major chromosome aberrations is brought in early because of its great social interest. Here, as elsewhere, I have tried to avoid excessive terminology and detail, which in my view could otherwise reach easily offensive levels. The basic facts of Mendelian, molecular, and cell genetics are introduced in Chapter Two via the sex (X) chromosome. This is relatively unusual but has at least one important precedent in the classical treatment by A. H. Sturtevant and G. W. Beadle, whose famous *Introduction to Genetics* appeared before World War II. This approach corresponds to some extent with historical development and may offer one of the easiest ways to grasp the chromosome theory of inheritance. Chapter Three introduces non-sex-chromosomes and classical Mendelian laws. As is well known, the study of human genetics differs from the study of other organisms, in which experiments are possible. Human genetics must be considered in terms of populations and of the matings that take place in these populations and supply material for genetic observation. Thus, analysis of populations and the Hardy-Weinberg rule is considered next, in the fourth chapter. In view of its central place for understanding human genetics, this rule is given considerable attention. Here it was inevitable to use some numbers, and even one or two formulas, but I hope it will be recognized that this was done discretely. Blood groups and most recent developments such as HLA; continuous variation and environmental effects, with special regard to behavioral traits such as IQ and schizophrenia; the interplay of mutation, selection, and drift in shaping populations and directing their evolution; hybrid vigor; and the loss of heterozygotes under inbreeding are the subjects of Chapters Five to Eight. The last chapter, Chapter Nine, carries the ambitious title of "The Future of Human Genetics" and considers with brevity some of the most tantalizing new developments,

hopes, and needs of research in this field and some of the contributions
they can make to human welfare.

By bringing so many things together in one short book, one can only introduce the essentials. It is my hope that the reader will be convinced that some knowledge of human genetics should be shared by all. I also hope that some of the readers will be stimulated to a further study of the subject.

L.L.C.-S.□

Introduction

Genetics started over a century ago with the famous crosses that G. Mendel made between varieties of garden peas. He recorded, for instance, the number of times the direct or the more remote progeny from a pink and a white flowered variety had pink or white flowers. By studying this and other common traits of these plants, he was able to make some important generalizations, which go under the name of Mendel's laws. Later research showed that these laws are valid in basically all plants and animals, including humans, who do not differ, in terms of basic genetic mechanisms, from the other living organisms.

The development of human genetics took place late, mostly because we cannot use in humans those breeding techniques which were so successful in studying experimental plants and animals. A late bloomer, human genetics soon caught up with the rest, however. Today one can learn all the basic facts of genetics using human examples. Some major genetic discoveries have been made on humans. Those experimental crosses which, only a few years ago, could not have been made on humans can now be made in test tubes, by crossing individual cells of different humans or even of humans and other animals.

It has been said that if Mendel's laws had not been discovered through studies of garden peas, they would have been discovered early

xiv   in this century through studies of the inheritance of blood groups or also of "inborn errors of metabolism." The inborn errors of metabolism are inherited defects or diseases in which abnormal substances, or the lack of a specific chemical reaction, can be identified as the primary factor of the defect. Thus, albinos cannot form the dark pigment that all other individuals, black or "white" (who are really not so white), can form and deposit in their skin. It is in great part from the amount of deposited pigment that the skin derives its color. It was postulated that the albino's inability to produce the pigment was due to an alteration of a gene that is normally responsible for manufacturing some factor essential for production of the pigment. This interpretation antedated by decades the experiments on fungi and bacteria which were designed specifically to test how genes act and which led to very similar conclusions. It was also in humans that, for the first time, the chemical nature of a specific gene was understood. Sickle-cell anemia is an inherited disease relatively common among people of African ancestry. Hemoglobin is a substance giving the red color to the blood; anemia is due to insufficient amounts of it. Hemoglobin is the carrier of oxygen from the air in the lung to the rest of the body. Insufficient oxygenation has serious, even mortal consequences. It has been known for some time that, in sickle-cell anemia, loss of hemoglobin occurs because the red cells in which it is contained break open under certain circumstances. But the primary problem is not in the walls of the red cells which break. An analysis of a physical property of hemoglobin (its mobility in an electric field) showed that the difference was in the hemoglobin molecule itself. The term "molecular disease" was thus coined; the expression "molecular genetics" was to follow. The peculiar electric behavior of the hemoglobin from sickle-cell anemics may be fully explained in chemical terms after more knowledge is accumulated on the chemistry of its molecule. Hemoglobin molecules are big, some 4000 times bigger than a molecule of water, yet their structure is practically fully elucidated today. One single molecule can be dissociated into several hundreds of simpler components. Hemoglobin from sickle-cell anemics differs from that of unaffected people in just a single one of these simple components. The nature of the alteration itself can explain the physical behavior of the molecule in an electric field, as well as the tendency of red cells to open up—thus losing hemoglobin and causing anemia in sickle-cell anemics. It is very unfortunate that in spite of all this knowledge there still is no simple treatment available for sickle-cell anemia. In other genetic disease, knowledge of the chemical mechanisms has led to a successful therapy. We can understand, however, why this therapeutic problem in sickle-cell anemics is a more formidable one than for diseases that are now under control.

Much of human genetics is based on the study of disease. The reason is simple. Interest in our health is so great that many efforts have already gone into this area. Since the turn of the century, the battle for health has been directed primarily against infectious diseases. Here, many important successes have been obtained, and mortality from such causes has decreased in advanced countries to almost imperceptible levels. The dramatic increase in life expectancy in the Western world in recent years is basically due to the introduction of antibacterial drugs. Although differences in resistance to some infectious diseases are in part genetic, here the primary factor of disease is just the bacterium. One can successfully eliminate it from the body of the diseased, using drugs that kill, or slow down the multiplication of, the bacteria without damaging the host. Once the bacterium can be eliminated, the problem will have disappeared or at least will have been greatly reduced in magnitude. But what diseases are left for the doctor, since bacteria are not as dangerous as they used to be? Among those that are left, the genetic ones emerge as being more and more important. And here are some truly ghastly diseases. Moreover, as our knowledge accumulates, their number increases, and we realize that diseases previously not recognized as such have a substantial genetic basis. Even an affliction like schizophrenia, a serious and frequent alteration of behavior which can have extreme consequences, has a genetic component undoubtedly of some importance.

Much human genetics is thus medical genetics, and if our interest were exclusively practical, leaving no room for intellectual curiosity, we might want to concentrate our efforts here. But what we learn from medical genetics is not only good for the patient and the doctor. The mechanisms of inheritance are the same, whether they lead to disease or to other types of variation that do not affect our health in a direct way. Thus we can learn genetics from the study of disease; but much knowledge has also accumulated on differences which are not in the realm of medicine. Most of the variations in human IQ fall within what is considered a "normal" range, and yet much attention has been dedicated to IQ, especially in the last few years. Unfortunately, we soon reach the limits of the power of modern genetic techniques; before we

xvi   can make some truly significant statements, we must wait for further developments of human genetics. But it is important that we realize what the present limits and future prospects are for this kind of analysis, lest we fall into facile generalizations and potentially wrong statements. "Environment"—the aggregate of factors which affect the development of an organism, besides its genes—is also powerful in shaping us. The interaction of genes and environment is complex, and we barely begin to understand its implications.

In many areas of genetics there is greater hope of obtaining results that are not only successful, but also truly useful and satisfactory. Some areas that emerge as being very challenging and potentially rewarding are, first, those of genetic engineering—the hope of correcting our "wrong" genes by "surgery" at the molecular level—and second, the study of environmental "mutagens." These are agents (physical and chemical) which can cause mutations, that is, genetic changes. Since mutations have a high chance of generating disease, they are clearly unwanted. But many mutagens still have to be identified. An interesting association strengthens our desire to eliminate or reduce the level of mutagens around us: most agents that induce mutations also cause cancer, and vice versa. Thus, human genetics has high relevance for human welfare, and is coming to take a greater and greater place in our basic knowledge.

SECOND EDITION
ELEMENTS OF
HUMAN GENETICS

Chapter one

A cell. Note external wall and the nucleus with a membrane. Inside the nucleus are a nucleolus and the chromosomes. This is an electron micrograph of a thin section. (From Watson, J.D.: Molecular biology of the gene, ed. 3, Menlo Park, California, 1976, W.A. Benjamin, Inc.; Courtesy Dr. K.R. Porter and Dr. M.C. Ledbetter, Biological Laboratories, Harvard University.)

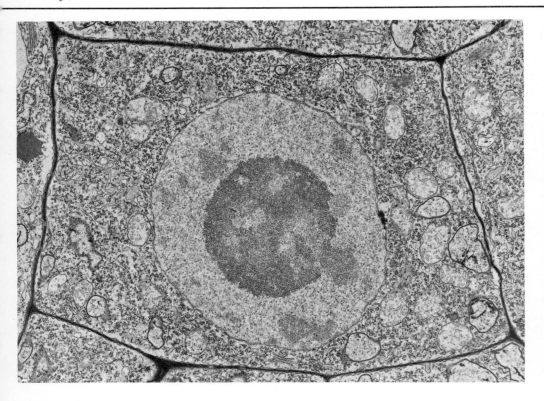

Chromosomes

*Cells and organisms have elaborate mechanisms making
it possible to produce copies of themselves that are practically
perfect. The relevant information is contained in the
chromosomes, and therefore chromosomes must be distributed
equally at reproduction of cells or organisms. Errors in
distribution of chromosomes have serious consequences.*

1.1 Cells and Chromosomes

That complex organism, the human individual, is made up of many
trillions of independent elements, which we may compare to Chinese
boxes, called *cells*. Most of these cells are different from one another,
and they often have considerably different functions, as exemplified
by the cells in the nervous, muscular, and all other tissues and organs
forming the human body. Inside each cell is usually another box, the
nucleus, which contains, in addition to other things, the *chromosomes*.
The chromosomes are the bearers of our genetic fate; in other words,
they determine how the cells are made and what we ourselves finally
become at the end of a very long period of development, which starts

2   when a sperm and an egg fuse to form a new individual and continues for a number of years.

In spite of the great variety of cells in the body, the chromosomes contained in them are practically always the same from cell to cell. There are always 46 of them (apart from a few exceptions, which we shall discuss later). Chromosomes consist mainly of a long thread of a substance called DNA (deoxyribonucleic acid). If the DNA thread forming a chromosome were entirely uncoiled, it would be, in the case of the average human chromosome, about two inches long on average. Since the width of the thread is two millionths of a millimeter, DNA can be seen only with the most powerful electron microscopes. Other substances, which help in determining the shape and function of chromosomes, are present in addition to DNA.

Fortunately for the ease of our observations, there are times when chromosomes tend to collect into bodies that are much thicker and shorter, that is, when the DNA thread is coiled in some complex, incompletely understood way, along with other substances. Each chromosome has then the form of a short body with some additional peculiarities (see Fig. 1.1). This shortening and thickening of the chromosomes (which makes them visible even with an ordinary microscope at a magnification of 1000 times) occurs during the period leading to division of the cell into two daughter cells.

As already mentioned, almost every cell in almost every individual has 46 chromosomes. Most quantitative assessments in biology are difficult, and therefore it is not astonishing that until about 20 years ago the number of chromosomes in man was believed to be 48. We now know that 48 is the number of chromosomes for gorillas and chimpanzees. Chromosome numbers are characteristic for each species, and they can vary from one to several hundreds.

Cells keep the number of chromosomes constant by a precise mechanism of subdivision of daughter chromosomes to each of the two daughter cells. Before cell division a DNA filament generates two identical chromosomes, and each daughter cell receives one of the two identical chromosomes, ensuring that one daughter cell is identical to the other daughter cell and that both are identical to the parent cell. This process of cell division, called *mitosis*, is shown in Fig. 1.2. Interphase is the process between one cell division and the other; only a round body inside the nucleus, the nucleolus, is clearly visible, while chromosomes are not. At the beginning of mitosis, they become clearly visible inside the nucleus (prophase); only two chro-

A chromosome. The constriction about the
middle is called a centromere.

A chromosome about to divide. The
centromere has not yet divided.

A chromosome after division of the
centromere. Two chromosomes identical
to the first one have formed.

Figure 1.1

mosomes are shown in the figure, for simplicity. Later the nuclear
membrane dissolves, a spindle of fibers appears, and chromosomes
align themselves in the middle of the spindle, to which they seem to
attach by their centromeres (metaphase). When centromeres divide,
they seem to be pulled to the poles of the spindle by the fibers, and
the daughter chromosomes start separating (anaphase). Towards the
end of mitosis (telophase), chromosomes of the daughter cells are at
opposite poles; membranes form again to shelter chromosomes in the
cell nuclei, and chromosomes once again become invisible. Two identical daughter cells have been generated.

1.2 Sex Chromosomes

The 46 chromosomes in each human cell are all in duplicate; thus we
really should speak of 23 pairs of chromosomes. With special techniques of observation of cells, we will find that these 23 pairs can to a
large extent be distinguished from one another. This is especially easy
today because of new staining techniques, by which we can also see
that the two members of each pair are practically identical to one another in most individuals. In the cells of a female individual all 23 pairs
are made of identical pair members, but in those of a male there is one
chromosome pair that is made of nonidentical members. We now
know that this pair of chromosomes is of special importance for the
determination of sex. In the males this pair is called XY, and the corresponding pair in the female is called XX. The X in the male is identical to the two X's in the female, and is a relatively long chromosome
(see Fig. 1.3); but the Y chromosome is a very short chromosome
indeed, and in a few rare individuals may even be missing.

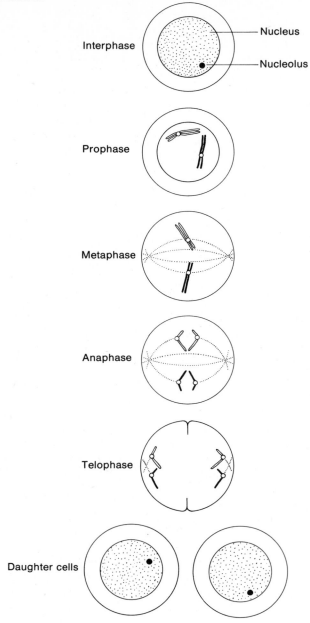

Figure 1.2
The schematic phases of mitosis. Only one pair of chromosomes is shown.

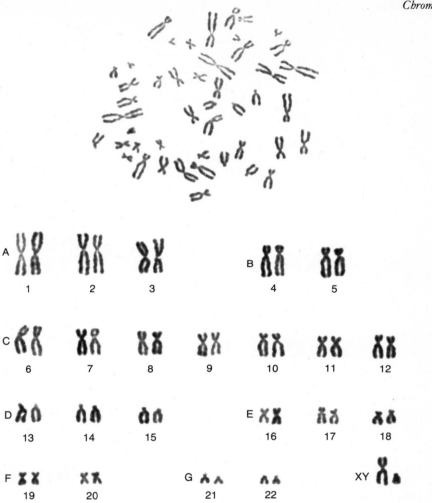

Figure 1.3
Above, human chromosomes from a lymphocyte (white blood cell) cultured in a test tube. The donor is a normal male. Below, the chromosomes photographed in the mitotic plate from a normal male shown above have been assembled in pairs, by pasting those which look identical (homologues) near each other and ordering the pairs according to decreasing size and ratio of shorter to longer arms. Not all 23 human chromosome pairs can be distinguished by this technique. Chromosome pairs which are difficult to distinguish one from the other are placed in a group with the same alphabet letter (from A to G)—in males, one chromosome pair is made of unequal members (XY). (Courtesy F. Nuzzo.)

(6)

Today we believe that the Y chromosome is essential for the development of masculinity, on the basis of observations such as the following.

1. There are some rare individuals who lack completely the Y chromosome and therefore have 45 chromosomes, only one of which is an X. They are called XO, where O refers to the fact that the Y chromosome is missing. The important aspect of this observation is that these individuals are females. They are not absolutely normal females, for they tend to be smaller, often have a webbed neck, and usually do not develop normal reproductive organs. This XO condition is also called Turner's syndrome, and is considered to be the result of an accident that occurred when the particular gamete (usually the sperm) that gave rise to the abnormal individual was formed.

2. On the other hand, there are other exceptional individuals who have more than 46 chromosomes. Those who have more than two X's (three, four, or even more) but no Y are almost normal females.

3. Further, there are some individuals with 47 chromosomes in whom there are clearly two normal X's, but also one Y. There is a clinical name for this condition, Klinefelter's syndrome, since there are some associated abnormalities. These individuals (XXY) are males. An individual is a male so long as he has at least one Y chromosome, even though he has two or more X chromosomes.

4. Another relatively rare type of individual (approximately one out of 1000 males) is called XYY, having two Y chromosomes. This individual is a practically normal male. He tends to be taller than the normal male by more or less the same amount an ordinary male is taller than an ordinary female. Because such individuals are found more frequently among prison inmates, suspicions have been voiced that XYY individuals may not be completely normal from a behavioral point of view, but the evidence thus far does not prove that such individuals should be considered socially dangerous. The possibility does remain that they may be somewhat more inclined, for instance, to petty theft and other misdemeanors, but this may also depend—perhaps even in its entirety—from the fact that XYY's seem to have substantially lower IQ's and may get caught more often when they misbehave. In any case, the whole problem is still under investigation.

In other words, however many X's an individual has, provided there is at least one X and no Y, that individual is a female. But if an individual has at least one Y and one X, that individual is a male (Table 1.1). Abnormalities of the X chromosomes thus suggest that the Y chromosome is responsible for determining the male sex, or, we may say, for switching the development of a fetus from being directed toward a female to being directed toward a male. The X chromosome, on the other hand, is perhaps not of great importance in the determination of sex, but as we shall see, it has a number of other known functions.

TABLE 1.1
Normal and abnormal karyotypes (chromosome sets) with respect to the sex chromosomes and sex of the carrier. In the last column, sex chromatin is indicated as explained in Section 2.9.

Chromosomes	Sex	Sex chromatin
XX	Female	1
XY	Male	0
XO (one X, no Y)	Female	0
XXX	Female	2
XXXX	Female	3
XXY	Male	1
XXXY	Male	2
XYY	Male	0

1.3 21-Trisomy or Down's Syndrome

Aberrations in the chromosomal constitution of individuals are not confined to the sex chromosomes. The first anomaly to be discovered concerned one of the smallest of all chromosomes, chromosome 21. The chromosome pairs are numbered in order of size, starting with the biggest, and the sex chromosomes are left out of the numbering; thus it is clear that chromosome 21 is the next to smallest. It was found that a peculiar clinical syndrome, once called mongolism and now commonly known as Down's syndrome, is usually accompanied by the presence of 47 chromosomes in the cells of the carriers of this condition. The extra chromosome belongs to type 21; therefore, this condition is sometimes called 21-trisomy. It might seem that one chromosome more or less should not be very important because we already have two of each type, but this is not so. A very delicate balance exists in the organism, and the production of a cell is an extremely complex process. The production of an organism containing trillions of cells is even more complex; any minor fault may lead to serious consequences. Individuals affected by 21-trisomy are usually recognizable as such at birth. Their development, especially intellectual, is delayed, they never develop into normal adults, and they tend to die relatively young. They are usually affectionate and quiet. Their facial expression is characteristic and the diagnosis is relatively easy to make. It

8   is claimed that these individuals can develop appreciably if they are given proper training, but such training is necessarily very costly.

A confirmation by analysis of the chromosomes is, in any case, useful because in some varieties of Down's syndrome there are 46 chromosomes. These are only apparent exceptions, for the individual really has three copies of chromosome 21: two normal copies, and at least one part of another chromosome 21, which has somehow become stuck to another chromosome. This is a consequence of an earlier breakage of a chromosome 21 and union of one of the segments with another chromosome; the process is called *translocation* (Fig. 1.4). Though a Down's child may have 46 chromosomes, its parent may be found to have 45 chromosomes and be clinically normal, because the parent does have two chromosomes of type 21, except that one of them (or most of it) is attached (translocated) to another chromosome. The discovery of such cases is important, because the frequency of trisomy in the children of individuals who have 45 chromosomes with a chromosome-21 translocation is fairly high, usually one in three. For all other individuals who are completely normal, the birth of a child with this type of trisomy is a rare event. However, it is not such a rare event if we realize that, on the average, one birth out of 700 does give rise to a 21-trisomy.

We do not know exactly why some individuals are born with 21-trisomy, but at least one fact is known: a disproportionately large fraction is born to women who are over 35 years of age at the time of conception. This means that when an older woman conceives a child, her eggs are more likely than those of a younger woman to have the abnormality described. The conclusion from the social point of view is clear. Women above 35 years of age should try to avoid conception, because they have a higher chance of having children with this major handicap. For women above 45 years this chance may be as high as 1 in 20. Today, however, it is possible, by a medical procedure (amniocentesis), to obtain, without damage to the fetus or the pregnant woman, fetal cells from which a diagnosis can be made of Down's syndrome and other pathological conditions. This can be done at a time when abortion is still possible. Not all religions and not all countries agree on the ethical and legal aspects of abortion in conditions like this. Also, serious genetic abnormalities, of which 21-trisomy is only one example, bring such anguish to the parents of the affected child, and are of such high cost to the family and to society, that it is to be hoped that religious attitudes will change accordingly. In recent years, in

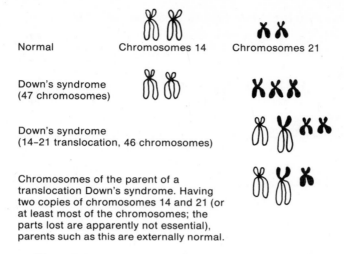

Normal
Chromosomes 14
Chromosomes 21

Down's syndrome
(47 chromosomes)

Down's syndrome
(14–21 translocation, 46 chromosomes)

Chromosomes of the parent of a
translocation Down's syndrome. Having
two copies of chromosomes 14 and 21 (or
at least most of the chromosomes; the
parts lost are apparently not essential),
parents such as this are externally normal.

Figure 1.4
Down's syndrome. Only chromosomes 14 and 21 represented.

fact, most advanced countries have changed or are changing their legislation to allow for abortion, on genetic and other grounds.

1.4 Other Chromosome Aberrations

There are many conditions in which trisomy of chromosomes other than 21 is found. Leukemia is a cancer of the blood system which is still difficult to cure and usually leads to death in a few years. In many cases of leukemia among patients who have a normal number of chromosomes, a peculiar modification of chromosome 22 is found. One of the two members of the pair is shortened and thickened and has a peculiar aspect. Such abnormal chromosomes found in the white blood cells of certain leukemia patients are called Philadelphia, after the place where they were first discovered. The finding indicates a correlation between that particular kind of leukemia and chromosome 22, although the relation is still not understood.

Other conditions involving trisomy of chromosomes other than 21 include absences (deletions) or duplications, even inversions of chromosome segments, as well as translocations (as already indicated), all of which give rise to a variety of clinical diseases. Most of these diseases are very serious but, fortunately, fairly rare. Altogether,

10   however, the frequency of clinical abnormality due to chromosome aberrations, including those of the sex chromosomes, is somewhat less than 1%. This is an evaluation made at birth. However, there surely are many more chromosome aberrations that have such serious consequences for the development of the fetus that they lead to death of the fetus in utero, that is, spontaneous abortion or stillbirth. Among the newborns that are found to have chromosome abnormalities, about one-quarter have 21-trisomy, one-twentieth have some other form of trisomy, another quarter have sex-chromosome variation, and the rest, about one-third of the total, have various rearrangements of the nonsexual chromosomes.

1.5 The Formation of Sperm and Eggs

We know that an individual is formed by the union of two cells, a sperm and an egg (Fig. 1.5). These cells are called *gametes*, male and female respectively. If sperm and eggs all contained 46 chromosomes, then from the fusion of one sperm and one egg would result a cell with 92 chromosomes. This obviously does not occur; therefore, some process of reduction must accompany the formation of gametes if the number of chromosomes is to remain constant from one generation to the next. This process is called *meiosis* (Fig. 1.6). Since the lack of a chromosome or the presence of an extra chromosome can, as we have seen, be very dangerous, the process of reduction must be very precise. It is essential to ensure that every individual contains the requisite 23 pairs of chromosomes. The requirement is satisfied if each individual receives one member of each pair from the mother and one member from the father. Therefore, when a gamete forms, be it sperm or egg, the members of each chromosome pair must separate in such a way that one particular sperm or egg cell receives only one member of each pair. It does not matter which, and therefore the probability that from a particular pair a sperm receives one or the other of the two chromosomes is exactly 50%. The same is true for an egg. Thus both sperm and eggs have 23 chromosomes each, one of each pair. When they fuse, in the process called fertilization, the cell that results will again have 46 chromosomes (Fig. 1.6). This resulting cell is called a *zygote*, and is ready to develop, through a process that takes many cell divisions and almost 20 years, into an adult organism.

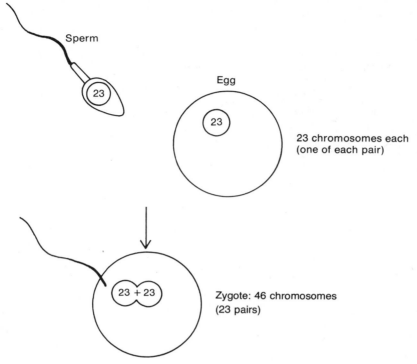

Figure 1.5
Fertilization. The number inside the nucleus indicates the number of chromosomes.

Meiosis is a modified form of mitosis, the usual process of cell replication, illustrated schematically in Fig. 1.6. It is important to note that meiosis is the time at which errors in the distribution of chromosomes are most likely to have serious consequences. Thus, if the two members of one particular chromosome pair do not separate at the time of formation of sperm, a sperm cell may result which has two members of one pair, and another that has no member of that pair. The chromosome aberrations that we encounter occur mostly because of errors in the distribution of chromosomes at meiosis.

Some individuals, however, are found, on close inspection, to have two types of cells, some having a normal chromosome set and others having an abnormal set. The usual explanation is that they have arisen because the chromosome aberration has taken place, not in the formation of gametes, but during the development of the individual during cell division in the growth process. These individuals

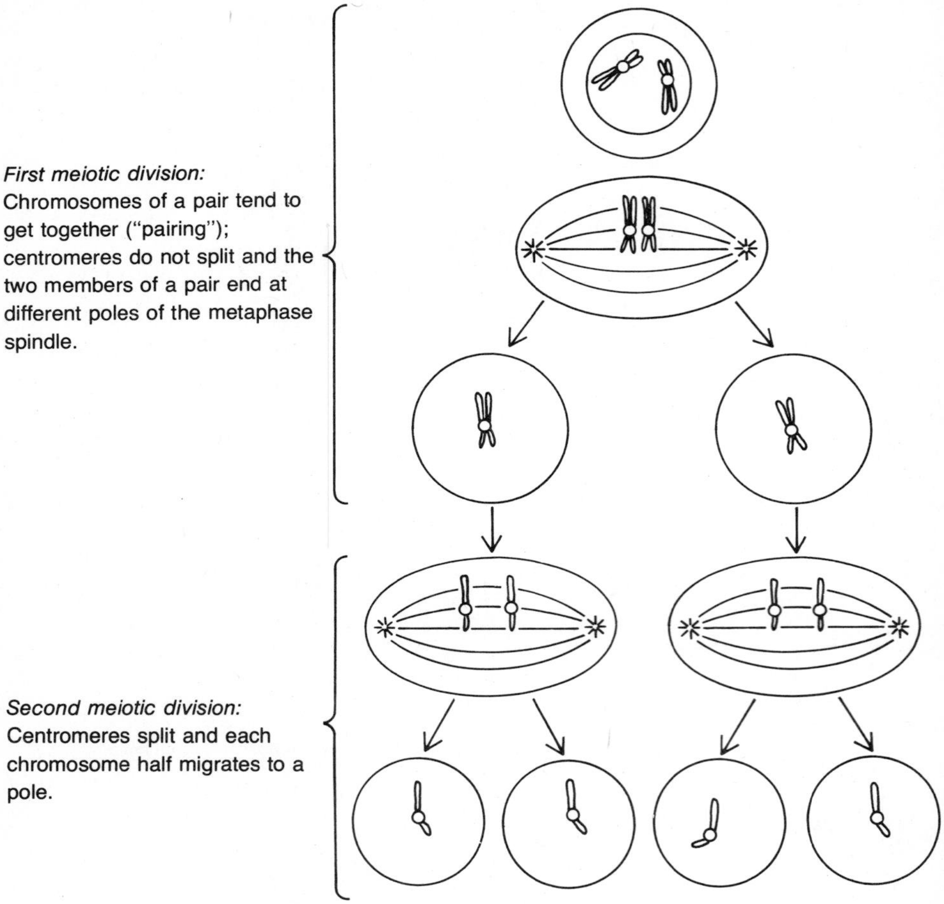

First meiotic division:
Chromosomes of a pair tend to
get together ("pairing");
centromeres do not split and the
two members of a pair end at
different poles of the metaphase
spindle.

Second meiotic division:
Centromeres split and each
chromosome half migrates to a
pole.

Figure 1.6
Meiosis. Schematically, only one chromosome pair is represented. Meiosis consists of two modified mitotic divisions. It produces four gametes each with reduced (23) number of chromosomes out of a cell with 23 pairs.

are called *mosaics*, and usually have conditions of lesser severity than do individuals in whom only abnormal cells exist.

We have seen the difference between males and females is in one chromosome pair: females are XX and males XY. When this chromosome pair is reduced at meiosis, gametes produced by females (eggs)

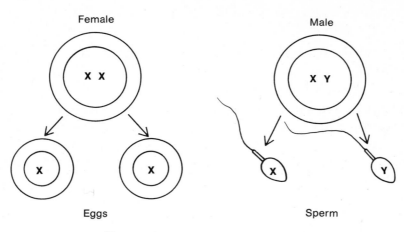

Figure 1.7
Formation of gametes and sex chromosomes.

can only contain an X chromosome. But gametes produced by males (sperm) can either contain an X or a Y (Fig. 1.7). Thus there is only one type of egg and two types of sperm, as distinguished by the sex chromosomes.

PROBLEMS

1. Are the following statements true as a rule for meiosis alone, for mitosis alone, for both, or for neither?

(a) Chromosome numbers are the same in parent cells and daughter cells.

(b) Chromosome numbers are halved during the process in a diploid organism, like man.

(c) At the end of the process there is only one member of each chromosome pair per cell.

(d) There are two members of each chromosome pair in each daughter cell (in a diploid).

14   ι   (e) A full description of the process requires the study of four cells.

2. What is the outward sex of (a) an XYY individual? (b) an XXXX individual? (c) an XXXY individual? State the rules you follow to make these predictions.

3. If an individual with Down's syndrome has 46 chromosomes, and has parents who look normal but whose chromosomes have not been tested, (a) what do you expect you might see in the chromosomes of one of the parents? (b) if this expectation is borne out, what are the chances that another child born of the same parents will have Down's syndrome?

4. What genetic danger do you foresee for a child born to a woman over 40 years of age?

Chapter two

Chromosomes of normal human female, from culture treated five hours before harvesting with radioactive precursor of DNA (tritiated thymidine); radioactivity is shown by small dark grains which have formed in a superimposed photographic film. X chromosomes are unique in many ways; only one of two X chromosomes in preparation has incorporated a substantial amount of radioactive thymidine, i.e. has synthesized DNA in period of experiment, as shown by grains around and above it. This is the same chromosome inactivated because of Lyon effect. Large dark blobs are nuclei not dividing at time of preparation. (Courtesy F. Nuzzo.)

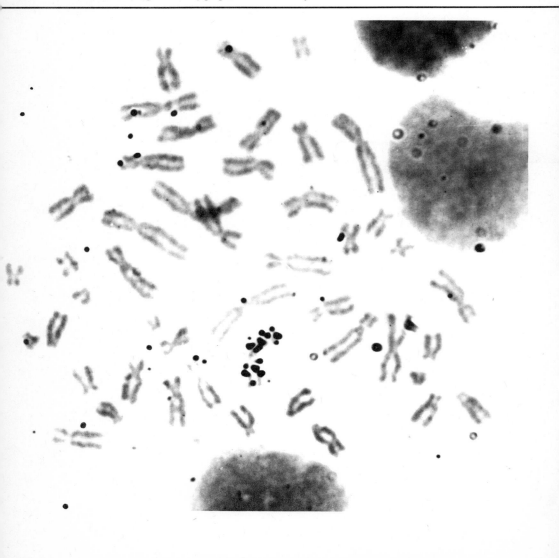

Sex and Sex-linked Inheritance

The X chromosome exists in only one copy in males and in two copies in females. This peculiar situation can give us a good introduction to the rules of inheritance and to the mode of action of genes.

2.1 The Numbers of Males and Females

We have seen that, with respect to sex chromosomes, all female gametes (eggs) are equal: they carry one X chromosome. But male gametes (sperm cells) are of two types: one type carries an X and the other type carries a Y. The two types of sperm should form in equal numbers, that is, half of the sperm cells should carry an X, and half should carry a Y. Thus the outcome of fertilization will be as in Fig. 2.1. We therefore expect exactly half the individuals conceived to be males and half to be females. We know exactly, however, only the sex ratio at birth, and it is probably similar to that at conception. There is, in fact, a slight prevalence of males (more than 51% of all births), for reasons that are not entirely clear. A slight excess of males may be useful for the human species because the death rate for males of all age groups

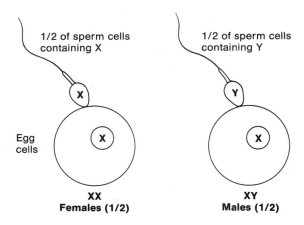

Figure 2.1
The formation of equal numbers of individuals of the two sexes at fertilization. Only the sex chromosome is indicated in each sperm and in the nucleus (the smaller circle) of each egg cell.

seems to be somewhat higher than that for females. At present there is an exact equality of males and females at about the age of sexual maturity, when equality in numbers is most convenient. Thus the excess of males at birth may represent an advantage to our species and be the consequence of adaptation by natural selection (see Chapter 8).

2.2 Inheritance of a Character Determined by the X Chromosome

Many substances present in cells are *proteins* (Fig. 2.2) which consist of one or more threads of simpler compounds called *amino acids*, and occasionally other substances as well. An *enzyme* is a protein, consisting usually of several hundred amino acids, which catalyzes (that is, permits the fast development of) a specific chemical reaction in the organism. There are many thousands of different enzymes, and each can perform a specific transformation of a very limited range of substances called *substrates* into other slightly different substances called *products*. For instance, the enzyme glucose 6-phosphate dehydrogenase (called, for short, G6PD) transforms glucose 6-phosphate (its substrate) into 6-phosphogluconate (its product) (Fig. 2.3). All enzymes are indicated by the suffix -ase, and each has a prefix or longer expression that identifies briefly its chemical action or its substrate or both.

Figure 2.2
Proteins are made of simpler substances, called amino acids, joined
together in long chains. Illustrated is a fragment of a protein.
In each circle is an amino acid (indicated by its name).

$$\text{Glucose 6 Phosphate} \xrightarrow{\text{G6PD}} \text{6-Phosphogluconate}$$
$$\text{(substrate)} \qquad\qquad\qquad\qquad \text{(product)}$$

Figure 2.3
Action of the enzyme glucose-6-phosphate-dehydrogenase (G6PD).

Figure 2.4 shows an example of an inherited difference in the enzyme. It is responsible for one step in the biochemical pathway which metabolizes the sugar glucose, and thus is involved in the utilization of the energy in sugar. It is contained in red cells and many others. In the figure, extracts of human red cells (containing the enzyme) have been subjected to electrophoresis, as described in the legend. Examination of the blood of many women has shown three types of interest with respect to this topic, as shown in Fig. 2.4. Some contain an enzyme form called A, some a form called B, and others contain both A and B. But among men only two types are found: only A or only B, but not both.

The simplest explanation that we can give for the fact that males cannot produce both types is that the enzyme G6PD is produced by the X chromosome. There must exist two G6PD enzymes (A and B) which are different enough that they migrate differently under electrophoresis, for reasons that we shall discuss later. Since a woman has two X chromosomes, it is possible that one makes enzyme A, while the other makes enzyme B. Thus a woman can conceivably have both enzymes (type 3 in Fig. 2.4). Another woman's X chromosomes may both be of the B-making type, so that she belongs to type 2 in the figure; or of the A-making type, so that she belongs to type 1. In men, however, the type with the two enzymes, A and B, cannot appear because men have only one X chromosome. We obviously assume that the Y chromosome is unable to make the enzyme.

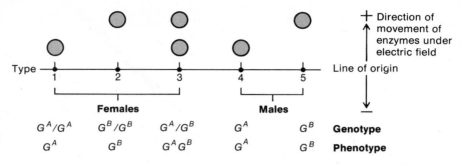

Figure 2.4
The blood of five different individuals (three females and two males) subjected to electrophoresis. At the beginning, a small amount of an extract of red cells from each individual is placed in a gelatinous plate at a point on the line marked "Origin" and an electric current is applied at either end of the plate. When the current is applied, the enzyme migrates, and the rate and direction of its movement are characteristic of the enzyme and the conditions of the environment (which we shall not discuss here). After a few hours the electric current is removed and the position of the enzyme is shown by a staining procedure which affects only the enzyme under study. Thus the spots in the figure represent the positions of the G6PD enzymes (see text) after electrophoretic migration. We see that the three females belong to three different types. One of them has a slow-moving enzyme (A), the second has a fast-moving enzyme (B), and the third has approximately equal quantities of both of these enzymes. The last two types on the right come from two males, one of whom has only the enzyme A and the other only the enzyme B. For definitions of genotype and phenotype, see Sections 2.3 and 2.7 respectively.

We shall see that this hypothetical explanation can be accepted with full confidence today, because there are many proofs that it is correct. We may wonder, however, why the two types of G6PD enzyme, A and B, migrate differently in an electric field. We know that a protein is made of many amino acids and that basically there are only twenty types of amino acids. Occasionally it is found that there are different forms of the protein with practically the same function; for instance, the forms of the G6PD enzyme are similar in function but differ from each other because a particular amino acid in a given position is substituted by another (Fig. 2.5).

Some of the 20 different amino acids forming a protein chain have an electric charge, which may be either positive or negative. These amino acids determine the direction and rate of movement of a protein in an electric field, a process called electrophoresis. When we find that two proteins like the A and B enzymes move differently under electrophoresis, we know that they differ in at least one amino acid. In one of the proteins one of the amino acids has an electric charge,

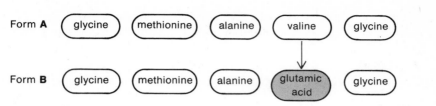

Figure 2.5
An amino acid substitution in two different forms of the same protein (only five amino acids represented).

while the corresponding amino acid in the other protein has no charge or has an opposite charge.

Why are there these two types of enzyme? We postulate that some change must have occurred sometime, possibly a very long time ago, in a segment of an X chromosome whose specific function was that of making the G6PD enzyme. We call such a segment with a specific function a *gene*. We do not know whether the original type was the A or the B form, but let us suppose that it was A. The change that turned the gene making enzyme A into that making enzyme B must be such that one amino acid only is substituted by another in the B form. We call it a *mutation*, and it may have occurred hundreds of thousands of years ago and been transmitted from parent to child down to us. The study of mutations is the study of evolution, of which we shall see more in this chapter and in Chapter 8. We are now concerned only with *predicting* what will happen with specific matings between individuals of the various types. We can only mate females with males and, as we have seen, every female gets one X chromosome from her father and one X chromosome from her mother. An individual of type 1 (see Fig. 2.4) who has only enzymes of type A, must have received both from her father and from her mother a chromosome in which the gene producing the enzyme G6PD was such that it produced the enzyme of type A. When we consider an individual of type 2, that is, a woman who has only the enzyme of type B, we repeat the same reasoning, simply changing A to B. With respect to the G6PD markers, therefore, both these individuals received the same gene from both parents, although type 1 received type A and type 2 received type B. Individuals who receive the same type of gene from both parents are called *homozygotes* (for that gene). Therefore, both type 1 and type 2 are homozygotes although they are different from each other. We use the name *alleles* for the alternative forms of the same gene (in the pres-

22 ent case, the two forms making the two enzymes A and B). We say that the two women are homozygous respectively for allele A and for allele B.

Clearly, the woman of type 3 has received a different gene from her father than from her mother: A from the father and B from the mother or vice versa (the result, in either case, is the same). Individuals who receive different allelic genes from the father and mother are called *heterozygotes* for that gene.

2.3 Predicting the Outcome of Matings

We now want to be able to predict the outcome of a mating between individuals whose composition we know in terms of the enzyme. We call *genotype* the genetic composition of an individual with respect to the gene or genes under discussion. Using the symbol* G^A for the gene making enzyme A and G^B for the gene making enzyme B, we can write the genotypes for the five individuals as follows: G^A/G^A, G^B/G^B, G^A/G^B, G^A, G^B. A slash is used here to denote alleles on a pair of chromosomes. Since there are three female genotypes and two male genotypes, altogether there are 3×2 (that is, 6) possible crosses. In order to predict the outcome of any cross, we always follow the same rules:

1. Specify the genotype of both parents.

2. Determine which gametes (sperm or egg) each parent can form and how many of each gamete type can be formed.

3. Combine the gametes of the father and those of the mother *at random* and obtain the proportion of each new zygote thus generated as the *product* of the proportions of the gametes that have formed the zygote. (A zygote, we recall, is the cell resulting from fertilization of an egg by a sperm.)

We will apply these rules now to one of the possible crosses. Let us imagine that a woman who is heterozygous, G^A/G^B, mates with a man who is G^B. The woman can form two types of egg, one G^A and the other G^B, and she forms them in equal proportions (half and half). The man can form only one type of X-carrying sperm, which is G^B. When this sperm meets the first type of egg, a G^A/G^B heterozygote is generated; and when the sperm meets the second, a homozygote

---

*For reasons of clarity, we do not use the standard symbols for genes, but simplified ones.

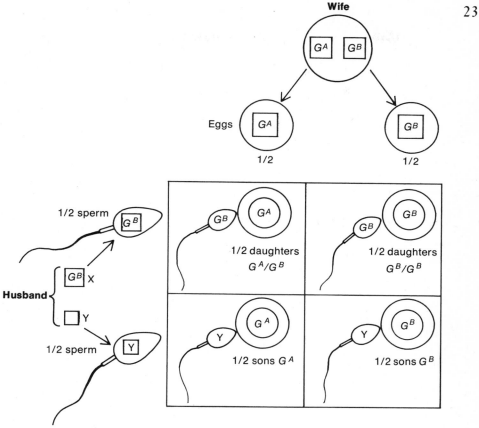

Figure 2.6
X-linked inheritance. Mating between a G^B *male and a heterozygous* G^A G^B *female.*

of the type G^B/G^B is generated (Fig. 2.6). This, of course, is what happens if the sperm carries an X, but we know that an equal number of sperm carry a Y, and such gametes will give rise to male offspring. Since the Y chromosome does not have the gene for this enzyme, the genotype of the male offspring will depend entirely on the gene contained in the particular egg that gave rise to him. In the cross we are considering, therefore, half of the male progeny will be of type G^A and half of type G^B.

The expectations of all six possible matings or crosses are given in Table 2.1. The cross we have described is in the bottom right box.

TABLE 2.1

Outcomes of the six possible crosses for G6PD enzymes. Crosses are considered for the two allelic forms G^A and G^B forming the A and B enzymes respectively. Each box represents possible children whose mother and father are given in the first column and row respectively. See also Fig. 2.4 for the method of detection of genotypes.

			Mothers		
			Type 1	Type 2	Type 3
	Mother's genotype		G^AG^A	G^BG^B	G^AG^B
	The eggs she forms		G^A	G^B	½ G^A and ½ G^B
Fathers	Type 4 G^A — Sperm types	X(G^A)	female children G^AG^A	female children G^AG^B	female children ½ are G^AG^A ½ are G^AG^B
		Y	male children G^A	male children G^B	male children ½ are G^A ½ are G^B
	Type 5 G^B — Sperm types	X(G^B)	female children G^AG^B	female children G^BG^B	female children ½ are G^AG^B ½ are G^BG^B
		Y	male children G^A	male children G^B	male children ½ are G^A ½ are G^B

The assumption we made at the beginning of this chapter, namely that the gene making G6PD is on the X chromosome (or, as we say, X-linked), is a plausible one, but what is the proof? The first piece of evidence comes from investigating the offspring of parents who exemplify the six matings in the table to see whether the expected proportions of the various genotypes are actually observed. This has been done in a great number of cases, and it has been shown that the predictions are completely accurate. This evidence clearly reinforces our idea that the G6PD is determined by a gene on the X chromosome. But this hypothesis fits nicely with a number of other known facts, and other independent proofs have been given.

2.4 From DNA to the Enzyme

The structure of DNA and the genetic code have by now app⸗ practically all newspapers and magazines; we shall give here only a simple summary of the whole story to refresh the mind of the student, who can find more details elsewhere. Chromosomes are chiefly threads of double-stranded DNA. Each strand is made of a long sequence of relatively simple compounds, the nucleotides, which are attached to one another. A nucleotide is made of the sugar deoxyribose, of phosphate, and of one of the following four bases: adenine, A; thymine, T; cytosine, C; and guanine, G. Since nucleotides are distinguished from one another only by the base, we shall identify the four different types by the letters A, T, C, and G, corresponding to the initial of each base. The two strands forming DNA are perfectly complementary (Fig. 2.7): whenever there is an A in one strand, in the other strand at the corresponding site there is a T; opposite T is A, opposite G is C, and opposite C is G. Knowing the sequence of nucleotides in one strand, we can predict the sequence in the other. When two new cells form from a single original cell, all the DNA in the original cell replicates (Fig. 2.8), and the process of chromosome separation at mitosis is such that the two daughter cells receive exactly the same chromosome set (46 each), and that for each chromosome in one of the two daughter cells there is an identical copy in the other cell.

A special mechanism in the cells enables proteins to be made from DNA by a process of "translation" from DNA to protein in accordance with a "dictionary" called the genetic code. But before this phase of translation there is another phase, transcription, in which one of the two strands of a DNA segment is "transcribed" (copied by simple complementarity rules) into a somewhat similar RNA (ribonucleic acid)

Containing just nine nucleotides: A T T A G A C A A G A T T A C

Its complementary strand: T A A T C T G T T C T A A T G

```
                          A T T A G A C A A G A T T A C
                          | | | | | | | | | | | | | | |
The double strand formed by their pairing:  T A A T C T G T T C T A A T G
```

Figure 2.7
A sample of a strand of DNA.

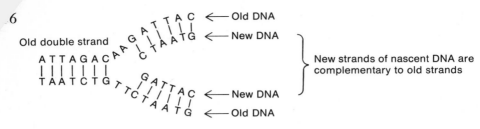

Figure 2.8
A double-stranded DNA in the process of duplication. Two double strands identical to the old one are formed, each made of half old, half new DNA.

strand called messenger RNA, which is single and complementary to the DNA strand being transcribed (see Fig. 2.9). The DNA segment transcribed corresponds to one gene or a set of neighboring genes, and includes thousands of nucleotides. Complementarity is the rule for transcription as for DNA copying, but in RNA there is another nucleotide U (for uracil), which takes the place of T and is therefore found opposite A in the DNA strand (Fig. 2.9). The process of making protein consists of building protein chains out of sequences of amino acids, chosen from among the 20 possible amino acids. The genetic code is such that three nucleotides in a row correspond to a given amino acid (see Table 2.2). Thus, the triplet in the DNA strand adenine-adenine-adenine (AAA) commands, according to the code, the introduction of the triplet uracil uracil uracil (UUU) in messenger RNA, and the introduction of phenylalanine (one of the 20 amino acids) in the protein chain. If the next three nucleotides in DNA are cytosine-adenine-guanine, then according to the code the next amino acid in the protein thread is always valine, and so on. Thus the sequence of nucleotides

Figure 2.9
A DNA strand and the messenger RNA being formed from it.

TABLE 2.2
The genetic code.

DNA	Messenger RNA	Amino acid	DNA	Messenger RNA	Amino acid
AAA	UUU	Phenylalanine	ATA	UAU	Tyrosine
AAG	UUC		ATG	UAC	
AAT	UUA		ATT	UAA	Chain end
AAC	UUG		ATC	UAG	
GAA	CUU	Leucine	GTA	CAU	Histidine
GAG	CUC		GTG	CAC	
GAT	CUA		GTT	CAA	Glutamine
GAC	CUG		GTC	CAG	
TAA	AUU	Isoleucine	TTA	AAU	Asparagine
TAG	AUC		TTG	AAC	
TAT	AUA		TTT	AAA	Lysine
TAC	AUG	Methionine	TTC	AAG	
CAA	GUU	Valine	CTA	GAU	Aspartic acid
CAG	GUC		CTG	GAC	
CAT	GUA		CTT	GAA	Glutamic acid
CAC	GUG		CTC	GAG	
AGA	UCU	Serine	ACA	UGU	Cysteine
AGG	UCC		ACG	UGC	
AGT	UCA		ACT	UGA	Chain end
AGC	UCG		ACC	UGG	Tryptophan
GGA	CCU	Proline	GCA	CGU	Arginine
GGG	CCC		GCG	CGC	
GGT	CCA		GCT	CGA	
GGC	CCG		GCC	CGG	
TGA	ACU	Threonine	TCA	AGU	Serine
TGG	ACC		TCG	AGC	
TGT	ACA		TCT	AGA	Arginine
TGC	ACG		TCC	AGG	
CGA	GCU	Alanine	CCA	GGU	Glycine
CGC	GCC		CCG	GGC	
CGT	GCA		CCT	GGA	
CGC	GCG		CCC	GGG	

A = adenine, C = cytosine, G = guanine, T = thymine, U = uracil

in DNA determines the sequence of amino acids in the protein, and the latter determines the properties of the protein. If a change occurs in DNA (for instance, if one of the nucleotides is accidentally replaced by another), the code determines if and how the amino acid that goes into the protein at that site is changed. This is an example of a *mutation*, and it may happen that the altered protein has different properties from the original one even if only one amino acid is changed. The change may be so great as to make the entire molecule nonfunctional, or less

28   functional, or to make it function in a substantially different way. On the other hand, the mutation may cause a difference that is slight or even unnoticeable. In the specific case of the two molecules G^A and G^B, the change does not seem to have affected the properties of the enzyme very much. The enzyme is functional in both forms, and there is perhaps no advantage to the individual in having one or the other of the two forms.

2.5 G6PD Deficiency

Other alleles of the G6PD gene exist which differ from the two discussed earlier in that they form enzymes that are very short-lived and consequently, in practice, not found in the cells. There are many different mutants of this kind, but a general symbol for them is g (or $G-$). Among individuals homozygous for this mutant, therefore, very little or no enzyme is found. They are thus unable (or less able) to metabolize glucose by the pathway used by individuals carrying the so-called normal allele. We use the letter $G+$ to identify the allele that makes the enzyme (A or B), and we call it "normal" simply because it is most common, but there is no reason to think that the capacity to make an enzyme is necessarily a part of normality. The concept of normality is entirely relative. The word normal may mean common or well functioning. The former definition, based on frequency, is the one most widely employed.

If we now consider simply the two alleles $G-$ and $G+$, the latter being A or B indifferently, we realize that we have exactly the same possibilities for inheritance that we had when considering A and B. Thus there will be three types of females, the two homozygotes (one positive and the other negative for the enzyme) and the heterozygote; and only two types of males (one positive and the other negative). Surprisingly, relative lack or even complete lack of this enzyme does not create too many problems for the carriers. However, individuals lacking this enzyme may have some complaints; certain drugs may cause in them severe hemolytic crises, which occasionally may even be fatal. In fact, the type $G-$ was discovered because it was found that several black American soldiers who were given primaquine, a drug used for the treatment of malaria, developed a hemolytic crisis, while the same phenomenon was much rarer among Caucasians. However, there are

Caucasian groups, for instance one Jewish group from Kurdistan, in which up to 60% of the males have been shown to be $G-$, further indication that this gene is not necessarily very dangerous to carriers.

2.6 Inactivation of One X Chromosome

G6PD deficiency has been especially useful in testing the hypothesis that, in any given cell of a woman, only one of her two X chromosomes is functional and the other is inactivated. This situation is often called *lyonization* after the name of its discoverer (M. Lyon). The inactivated X chromosome is one member of the pair in about half of all cells, and the other member in the rest of the cells. This has been especially easy to show with the enzyme G6PD because it is possible to stain each red blood cell differently according to whether it contains the enzyme or not. Investigation has shown that about half the red cells of a woman who is heterozygote *Gg* contain the enzyme while the others do not. This fact was used to make several elegant observations. One of them, the experiment illustrated in Fig. 2.10, shows that malarial parasites cannot prosper in red cells that do not contain the enzyme. The reason may be that the parasites need a particular substance, which is partially or severely depleted when the enzyme G6PD is lacking. Thus the malarial parasites do not thrive well in individuals all or part of whose red cells are partially devoid of the G6PD enzyme. This reduced susceptibility to malaria may compensate for the enzyme deficit which these individuals ($G-$) show. Thus natural selection (see Chapter 7) may favor $G-$ genes in areas where malaria is strong. It should be remembered that malaria is one of the worst killers in tropical countries and in some temperate areas. Therefore, the presence of many $G-$ genes is an indication that malaria is or was a major disease in the area of origin of the people under study.

2.7 A Deleterious Recessive: Hemophilia

G6PD deficiency is a case in which natural selection favors an allele in some environments but not in others. There may be mutants of other genes which are favored in all environments; the frequency of these mutants will therefore increase everywhere. There are, on the other hand, many cases in which the mutant allele is always disadvan-

Figure 2.10

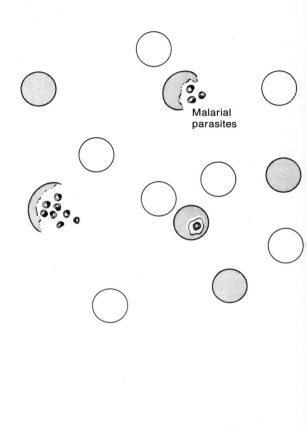

Malarial parasites

*This drawing represents a slide of blood taken from a woman who is heterozygous for the genes G+ and G−, also called Gg, and who is undergoing a malarial attack. Because of the inactivation of one X chromosome in each cell (lyonization), half of the red cells in which the G+ carrying chromosome is active produce the enzyme G6PD, while the other half (in which the G− chromosome is active and therefore no, or only little, G6PD enzyme is found) do not. Those that contain G6PD are stained dark, and those that do not contain it are not stained. The smaller organisms are malarial parasites (highly diagrammatic) and they are found to multiply well only on the former type of cell. The drawing is based on a free interpretation of an experiment by L. Luzzatto, E. A. Usanga, and S. Reddy (*Science, 1969, Vol. 164, p. 839*). In actual experiments parasitization of G− cells does take place, but at lower rates than that of G+ cells.*

tageous in comparison to the normal. It is not difficult to understand that mutant alleles which are disadvantageous should tend to remain at a low frequency in the population, for natural selection is against them (see Chapter 7 for an explanation of why they persist at all).

A famous example of a deleterious gene located on the X chromosome is the gene that causes the disease called hemophilia. There are at least two types of this disease; the more severe is called hemophilia A, and the other, less severe, hemophilia B. They are both rare and their total frequency is of the order of one in ten thousand. Hemophiliacs are, or at least were, all males and we will easily understand why. Hemophilia is a disease in which blood clotting is retarded because of lack of a protein normally present in blood. A hemophiliac who has a wound may die from loss of blood because the wound will not stop bleeding. Today we can save these individuals by giving them the necessary protein taken from the blood of a normal individual, but until a short time ago this therapy did not exist, and most hemophiliacs died before reproducing and were soon eliminated from the population. Hence, the frequency of hemophiliacs has always remained low. Again, as for G6PD, there can be three types of females, one of them being heterozygote *Hh* (if we call *H* the normal allele and *h* the hemophilia allele, which cannot produce the specific protein necessary for initiation of blood clotting). In heterozygous women one chromosome is capable of producing the normal protein and the other is not. It turns out that in these women about half the amount of protein found in the normal homozygote female is present, but this is enough for normal functioning. Therefore, a heterozygous woman is not a hemophiliac.

We predict the inheritance of hemophilia in Table 2.3, which is copied from Table 2.1 substituting *H* for G^A and *h* for G^B. Thus, if a woman is heterozygous for hemophilia (*Hh*, last column), half of her male children will be hemophilic (*h*) and half will be normal (*H*). Since there are so few hemophilic males, the husband of this woman will probably be normal (*H*). Half of her daughters will be normal carriers of the gene like herself (*Hh*), and half will be completely normal (*HH*). For a woman to be hemophilic, she must be homozygous for the gene that cannot make the protein, and for this to happen, she must be the daughter of a hemophilic man and a heterozygous (*Hh*) or homozygous (*hh*, or hemophilic) mother. Until recently, hemophilic men were usually unable to reach the age of reproduction, and therefore hemophilic women were extremely rare. It is thus not surprising that almost all hemophiliacs known today are males. This reasoning can be made more precise, as we shall see later.

In a heterozygote, an allele that is not effective in the presence of another allele is called *recessive*. An allele that *is* effective in the heterozygote is called *dominant*. In the case of hemophilia, therefore, *H* is dominant and *h* is recessive. Because of the phenomenon of dominance, we must distinguish the genotype of an individual from his *phenotype*, which is what we can detect by a given method of observation. Judging on the basis of abnormal clotting, we cannot clearly distinguish a heterozygous woman, *Hh*, from a homozygote normal, *HH*. Their phenotype, as determined by the capacity to clot normally or almost normally, is the same, but their genotypes are different. More sophis-

TABLE 2.3
Inheritance of hemophilia—this table is built like Table 2.1. Because of recessiveness of hemophilia, the disease appears only in homozygous women *hh*, or in *h* males.

		Mothers		
		Type 1	Type 2	Type 3
	Mother's genotype	*HH*	*hh*	*Hh*
	The eggs she forms	*H*	*h*	½ *H* and ½ *h*
Type 4 normal — Sperm types X(*H*) / Y (*H*)		female children *HH* male children *H*	female children *Hh* male children *h*	female children ½ are *HH* ½ are *Hh* male children ½ are *H* ½ are *h*
Type 5 hemophiliac — Sperm types X(*h*) / Y (*h*)		female children *Hh* male children *H*	female children *hh* male children *h*	female children ½ are *Hh* ½ are *hh* male children ½ are *H* ½ are *h*

(Fathers — left margin label)

ticated laboratory tests may show the difference. In the case of G^A and G^B we do not speak of dominance or recessiveness, because both alleles are effective in the heterozygote; instead, we speak of *codominance*.

Color-blindness is another classical case of sex-linked inheritance which we should at least briefly mention. There are several types, most of which are due to genes located on the X chromosome. They involve the incapacity to produce one or more of the pigments in the retina that are necessary for the perception of colors. As a consequence, color-blind people are unable to distinguish certain colors. To classify the various types of color-blindness, the subject is asked to distinguish various figures or numbers which are printed with different colors and can be perceived entirely only by fully normal individuals.

The frequency of color-blindness varies with populations and is highest among Caucasians, where it can reach the surprising figure of 8% (this figure includes all types of color-blindness) among males. Among females the frequency is much lower because the normal alleles are dominant, and therefore heterozygous women behave, as in the case of hemophilia or G6PD, like normal people with respect to color-blindness (see also Chapter 5 on this point).

2.8 The Map of the X Chromosome

There are more than 100 conditions, mostly rare, caused by genes known—on the basis of observations similar to those described above—to be on the X chromosome. Each of these conditions is due to a mutation in a particular gene—that is, a segment of the chromosome having a specific function, usually that of making a specific protein. There may be genes somewhat more complex than this simple definition implies, in that they include several adjacent units with similar or related functions. Since we know (1) the average size of a protein, (2) the relative length of the X chromosome with respect to the other chromosomes, and (3) the total amount of DNA contained in the cell, we can estimate approximately how many genes there are in the X chromosome. There should be at most 200,000 genes. This is likely to be an overestimate; in any case, there are likely to be at least 5,000 or 10,000 genes on an X chromosome, showing that our knowledge is still extremely limited.

It may be interesting to mention how these numbers are derived. The total amount of DNA in one gamete cell is four billion pairs of nucleotides. (We count "pairs" because the DNA thread is double and the two strands are perfectly complementary.) Only one of the two strands is effective in making the messenger, and therefore we can consider from this point of view that the pair acts as a unit. A single X chromosome is about 5% of the whole ensemble of 23 chromosomes (sometimes called the genome). Thus, there should be about 200 million nucleotide pairs in the X chromosome. If the average protein is made of 300 amino acids, the DNA length that corresponds to it is about 900 nucleotides, because three nucleotides correspond to one amino acid. The total of 200 million nucleotide pairs on the X chromosome, divided by the number of nucleotide pairs corresponding to a protein and therefore to the gene making it, amounts to a little over 200,000; this is an estimate of the maximum number of genes

34   that may exist on the X chromosome. It is likely, however, that some part of the DNA is repetitious or does not carry genetic information, or is destined for other functions or for making types of RNA that are not translated into proteins. Hence, the actual number of genes that specifies proteins may be smaller than this estimate—according to some, even as little as 2–5% of it. This explains the minimum estimate of the number of genes on the X chromosome given previously.

It is clear that if the X chromosome were passed as an entire unit from parent to child, all the genes that are on one X chromosome would tend to stay together and would be passed as a block. But at the time of gamete formation there is a phenomenon called *crossing-over*, which permits a reciprocal exchange of parts between the two X chromosomes. This exchange is perfectly complementary, as shown diagrammatically in Fig. 2.11. Crossing-over is not limited to the X chromosome but happens also in all other chromosomes, as we shall see. However, it is easier to study in the X chromosome because we can immediately see its consequences in the male progeny of a heterozygous female.

Figure 2.12 shows the expectations for the male progeny of a female who has on one X chromosome the *G* and the *C* genes she received

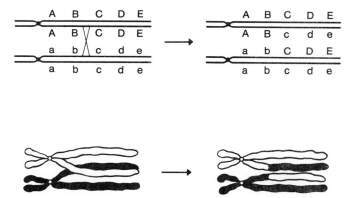

Figure 2.11
The X chromosomes undergoing crossing-over. The letters indicate hypothetical genes. Note that crossing-over actually takes place at a stage in meiosis when chromosomes have duplicated and centromeres have not. Capital and small letters are used to distinguish the two chromosomes which pair at meiosis. As each chromosome has duplicated, identical letters are used for the duplicates of a chromosome.

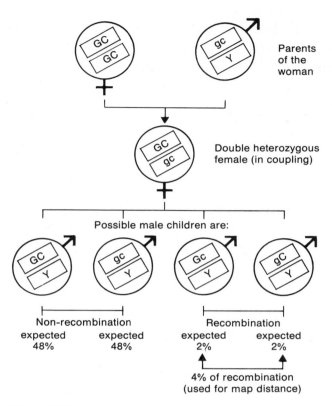

Figure 2.12
Ideal pedigree of a woman who is doubly heterozygous (in coupling) for color blindness (Cc) and G6PD deficiency (Gg). Capital letters indicate normal alleles, which are dominant. Bars inside ♂ (male) or ♀ (female) symbols indicate chromosomes. The parents of the woman are also indicated.

from her mother, and on the other the recessive *g* and *c* genes. Thus she is a double heterozygote, but she is phenotypically normal for these two genes. If there were no crossing over, she could have only two types of eggs in equal numbers: those carrying the chromosome she got from her mother, having both normal genes for color-blindness and G6PD (*C,G*), and those carrying both recessive alleles, which she received from her father (*c g*). These types are called parentals. However, because crossing over takes place, a proportion of gametes will

36   form in which the genes are exchanged. Thus, altogether four types of gametes are expected and the proportions in which they will be formed are indicated in the figure. Two new chromosomes are formed which are called recombinants; one of these has the *G* and *c* genes, and its complementary type has *g* and *C*. They are both expected in equal numbers, and are called *recombinants* to contrast them with the *parental* types of gametes. The percentage of recombinant gametes formed over all gametes is called the *percentage of recombination*. It is easy to count recombinants in sex-linked inheritance by looking at the male progeny of a woman who is heterozygous for two genes on the X chromosome. We will find there the X chromosomes of the mother without having to worry about the possible gametic contribution of her husband, who gave to his sons only a Y chromosome. The Y chromosome does not seem to have any influence on characters determined by genes found in the X chromosome.

The pedigree shown in Fig. 2.12 indicates an ideal family in which various cases of crossing-over and non-crossing-over have occurred. By studying many real families and summing the data in suitable ways, one can determine the actual percentage of recombination. Note that there are two possible types of double heterozygote (see the pedigree). In one, both dominant alleles are on one chromosome and both recessive alleles on the other. These are referred to as "in coupling" or *cis*. The other double heterozygote has one dominant and one recessive allele on each of the two chromosomes and is called "in repulsion" or *trans*. The woman in Fig. 2.12 is a double heterozygote in coupling (*GC/gc*) where the bar separates the two X chromosomes. She got one chromosome (*GC*) from her mother and the other (*gc*) from her father. Most of her sons will receive either *GC* or the *gc* chromosome (which are called parentals, being those she got from her parents). A few of her children will obtain *recombinant* chromosomes, *Gc* and *gC*, formed because of crossing-over between the woman's two X chromosomes. If she were a double heterozygote in repulsion (*Gc/gC*), the expectations for her male progeny would simply be exchanged, nonrecombinants becoming recombinants and vice versa; that is, we would expect 2% of each of the *GC* and *gc* types, and 48% of the *Gc* and *gC* types.

A very extensive study on animals, plants, and microorganisms has made it clear that the percentage of recombination between two genes depends on how far apart they are on the chromosome. The

closer they are, the less recombination there is; the further apart, the
more recombination. Therefore, the percentage of recombination is
a simple measure, though not a perfect one, of the distance between
two genes on the chromosome. This makes it possible to order the
genes on chromosomes and to construct maps of distances, given as
percentages of recombination. Figure 2.13 shows a map of the X chromosome including only a few of the genes known. Certainly within
the next few years our knowledge of recombination will be greatly
enriched.

2.9 Sex Chromatin

To see human chromosomes in the laboratory we have to take cells,
usually white blood cells, and let them divide by cultivating them in
test tubes. In nondividing ("resting") cells, chromosomes stain in the
nucleus and show up as a rather irregular net. It has been discovered
that many resting cells have one small DNA blob within the nucleus

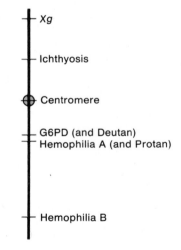

Figure 2.13
*A map of the X chromosome showing a few genes
whose position is known with greater than
usual accuracy. Xg is an X-linked blood group.
Ichthyosis is a skin disease. Deutan and protan are
forms of color-blindness.*

a b c

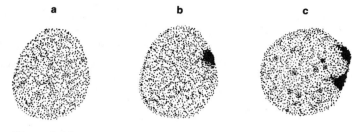

Figure 2.14
Sex chromatin (or Barr body). The nucleus of a cell with (a) no sex chromatin (usually, a male), (b) one Barr body (usually, a female), and (c) two Barr bodies (as found in individuals with three X chromosomes). In general the number of Barr bodies is equal to the number of X chromosomes minus one.

(Fig. 2.14). This blob is usually found only in females and is called *sex chromatin*, or Barr body. In fact, one can distinguish the sex of an individual by this kind of test, which can be carried out, for instance, on cells scraped from inside the cheeks.

The hypothesis has been advanced that sex chromatin is an inactivated X chromosome, or at least that part of it which is inactivated (perhaps not all of it is). We have already mentioned the phenomenon known as lyonization. Some further evidence comes from the analysis of individuals who have abnormalities in their sex chromosomes. Thus an XXY individual, who is, as we have seen, an almost normal male, has the sex chromatin of a female because one of the two X's is inactivated. On the other hand, XXX is an almost normal female, and has two rather than one sex chromatin blobs, and so on. In general, sex chromatin is an inactivated X chromosome, and all X chromosomes of a cell are inactivated in somatic cells except one. This is the rule by which one can predict the number of Barr bodies (sex chromatin) in the X chromosome karyotypes listed in Table 1.1.

2.10 Y Chromosomes

In the X chromosome we have found no genes that have to do with sex. We can hypothesize that the Y chromosome, when it is present, changes a potential female into a male and therefore must have "genes for masculinity." Not much more is known about the Y chromosome, however. Very few genes have been identified on it which are not concerned with sex. One of them determines the capacity to grow hair in the ear. Because this gene is contained in the Y chromosome and the Y chromosome makes an individual a male, it is clear that this character will be found only in males. It is not, however, found in all males, but it is expected that all male children of males who have this trait will also have the trait. Other Y-linked genes (as they are called) have been described, but are now somewhat in discredit.

PROBLEMS

1. Which is the sex that produces gametes of a different kind with respect to the sex chromosome?

2. A woman heterozygous for the X chromosome recessive gene determining the disease ichthyosis (a skin disease) is married to a normal male. (a) Is the woman affected by the disease? (b) Will her daughters be affected? (c) May she have affected sons, and if so, how many? (d) Will her normal sons transmit the disease? (e) Will her normal daughters transmit the disease?

3. A man has the rare dominant X chromosome-determined disease "vitamin D resistant rickets." He is married to a normal woman. (a) Will the sons show the disease? (b) Will the daughters? (c) If affected daughters have the disease, to which of their progeny will they transmit it?

4. A color-blind woman and a non-color-blind man have an XXY child who is not color-blind. (a) Does this tell us whether the chromosome aberration occurred in the sperm or in the egg that formed the XXY child, given that color-blindness is recessive? (b) What conclusion can we reach if the father was color-blind and the mother normal? (c) What conclusions can we reach if, with parents as in (a), the XXY child were color-blind?

5. What amount of antihemophilic globulin (the protein produced by the *H* gene) do we expect in *H* males, compared with that found in *HH* and in *Hh* females?

Chapter three

In earlier figures of dividing human cells, routine staining of chromosomes was employed which does not allow one to distinguish all chromosome pairs. A better resolution is obtained by heat treatment preliminary to staining which determines a banded appearance characteristic of each chromosome pair. This method is called G-banding—the same chromosomes are shown paired in this chapter (see Fig. 3.6). (Courtesy F. Nuzzo.)

Genes on Other Chromosomes

Non-sexual chromosomes carry the great majority of genetic material. It is to chromosome behavior that Mendel's laws owe their validity. These laws help us predict the progeny expected in a specified mating.

3.1 Autosomes and Mendelian Inheritance

We have started by studying one special type of inheritance, the so-called sex-linked one. But the majority of genes are on other chromosomes. The X chromosome constitutes only about 5% of the whole genome and the Y seems to be almost devoid of genes apart from those that determine maleness.

The chromosomes other than X and Y are called *autosomes*. The rules of inheritance for autosomal genes were discovered a long time ago by Gregor Mendel, who published the results of his investigations in 1865. Unfortunately, Mendel's paper was not understood by his contemporaries, or it escaped their attention. It was only in 1900 that three different scientists rediscovered Mendel's laws. At the time

42   Mendel made his discoveries, chromosomes and their behavior during cell division and fertilization were not known, and it seems a miracle that without this knowledge he could have arrived at those simple ratios that he explained so intelligently. Once chromosomes were known, and the chromosomal theory of inheritance had been put forward, it became much easier to understand the rules for the inheritance of autosomal genes.

3.2 The Disease Galactosemia

To illustrate these rules, we shall take as an example a rare disease, galactosemia. One may wonder why, here and elsewhere, we use mostly traits that are very unusual. Why not use eye color, nose shape, facial expression, etc.? The answer is that many of these latter characters have a complicated inheritance, that is, they may be determined by many genes. Therefore, they would not be suitable examples for understanding the basic rules of genetics; obviously it is simpler to study one gene at a time. This strategy, incidentally, was one of the secrets of Mendel's success.

Fortunately, very few children are born galactosemic. The main problem of galactosemics is that they lack an enzyme (galactose 1-phosphate-uridyl transferase) necessary for metabolizing the sugar they receive in the mother's milk. The absence of the enzyme is sufficient to cause in the children a very unfavorable and dangerous reaction when they are fed ordinary milk. We need only substitute a different sugar for the one in the milk to make these children healthy; but if this is not done, the damage may be serious, perhaps resulting in mental deficiency or even death.

Children that come under our observation for this disease usually have normal parents, and frequently there is no history of the disease among their relatives. A child with the disease is often born after other normal children. What makes us think the disease may be inherited is that more than one child born to the same parents may show the disease, which is otherwise extremely rare. Can we make more exact predictions? If the disease is due to a defective gene, incapable of— or markedly deficient in—producing the enzyme, we can draw exact expectations for these events. Because each autosomal pair has two members, for each autosomal gene there will be three genotypes, provided that two forms of the gene are known. Calling the two allelic forms of this gene gal+ and gal−, we will refer to the allele that cannot form the normal enzyme as gal−. Individuals can therefore be of three types: gal+ gal+ (homozygous for the normal gene), gal+ gal− (heterozygous), or gal− gal− (homozygous for the abnormal gene).

When we measure the quantity of the enzyme galactose 1-phosphate-uridyl transferase produced by individuals, we find that homozygous normals manufacture twice as much enzyme as do heterozygotes. This shows that each of the two chromosomes of the pair works independently. The amount of enzyme produced by the heterozygote, although half the normal level, is sufficient for normal functioning. Thus the heterozygote (gal+ gal−) does not suffer from galactosemia, and is indistinguishable from homozygous normals (gal+ gal+) unless the enzyme assay is done. Homozygotes for the gal− allele produce either no enzyme, or a very small amount.

In Fig. 3.1 we show what happens in a mating between two heterozygotes. The circle indicating each individual is divided into two halves; black areas represent the normal allele of the gene for the galactose enzyme, while white areas represent the abnormal, galactosemic allele. Because both parents are heterozygotes, each is represented by a circle that is half white and half black. To find what proportion of their children will be galactosemic, we use the same procedure that we followed in Section 2. First we find out what the gametes of each parent are, and how many of each we can expect. We then make random combinations of the possible gametes and, as the final step, we count the progeny.

It should be easy to understand Fig. 3.1 if you remember (1) that gametes, whether sperm or eggs, are either white or black, but not white *and* black, because they contain only one of the two chromosomes; and (2) that the two types of gamete (white or black) are formed in equal proportions by a heterozygote. It is also useful to remember that to compute the probability of union of one particular type of sperm and one particular type of gamete, one simply multiplies the relative proportions. Thus the proportions of zygotes given in Step 2 of Fig. 3.1 are computed simply as the products of the proportions (that is, the fractions) of the corresponding gametes. In this simple example, they all turn out to be $\frac{1}{2} \times \frac{1}{2} = \frac{1}{4}$. As shown in Step 3, we can expect one-quarter of the children to be normal, half to be heterozygote, and one-quarter galactosemic. We are speaking here of the genotypes of the progeny. If we were to look at the phenotypes of the progeny without making an enzyme assay, we could distinguish only

Father and mother are *both*
heterozygotes for galactosemia

Children?

Step 1

Gametes produced by

FATHER: 1/2   1/2

MOTHER: 1/2   ○   1/2   ●

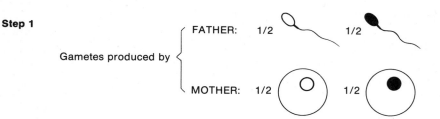

Step 2   Random combination of gametes

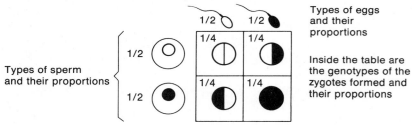

Types of eggs
and their
proportions

Types of sperm
and their proportions

Inside the table are
the genotypes of the
zygotes formed and
their proportions

Step 3   Counting types expected in the progeny

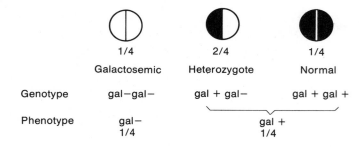

	1/4	2/4	1/4
	Galactosemic	Heterozygote	Normal
Genotype	gal−gal−	gal + gal−	gal + gal +
Phenotype	gal− 1/4		gal + 1/4

Figure 3.1
A mating between heterozygotes for galactosemia. The black areas in the circles stand for the normal
(gal+) gene, the white for the galactosemia (gal−) gene.

between normal and galactosemic; we would find three-quarters normal and one-quarter galactosemic (Fig. 3.1).

These are the standard proportions expected in a cross between two heterozygotes. They are often expressed as 1:2:1 for homozygote normal : heterozygote : homozygote abnormal, and also as 3:1 (3 for the dominant phenotype and 1 for the recessive phenotype). The cross we have indicated in Fig. 3.1 is only one of six possible crosses, but it is the most complicated of them all. With an understanding of Fig. 3.1 it should be easy to follow the consequences of all the types of crosses listed in Table 3.1.

3.3 Mimic Genes

In this section we shall briefly outline some possible complications in inheritance that are instructive. For example, there is a rare inherited trait called *albinism*, which is the complete absence of pigment in any part of the body. Albinos are not simply light-skinned, they are so white that they do not even have pigment in their irises. They have pink or very pale eyes, and white hair.

The main pigment in the body, called melanin, is formed by a series of reactions starting from the amino acid tyrosine (Fig. 3.2). If there is a block in any of the steps of the series, no pigment will be produced. Each step is controlled by an enzyme and each enzyme by a gene, that is, a particular chromosome segment. We shall use the term *locus* (plural *loci*) to refer to a gene in this sense. A locus usually makes one specific protein (or polypeptide chain that is a part of a protein). The use of this term will make it clear that we are speaking of two different genes in the sense of two different chromosome segments that make different proteins or polypeptides, as contrasted with two alleles of one gene, which are alternative forms of one gene or locus. Two alleles usually differ by only one mutation which may have changed only one amino acid in the protein. But two different loci may produce proteins that are widely different. If these proteins are, as in the case of albinism, enzymes involved in the production of melanin, lack of one enzyme or the other will cause melanin formation to cease.

There are, in fact, different types of albinos, differing in the locus that is mutated, and therefore in the metabolic step that is affected;

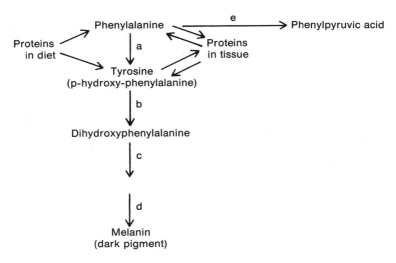

Figure 3.2

Some of the steps of the metabolic pathways affecting the two amino acids phenylalanine and tyrosine. They are both present in the diet, and in the tissue proteins. The human organism can convert phenylalanine into tyrosine (arrow a) by an enzyme called phenylalanine hydroxylase. Absence of this enzyme is responsible for the mental deficiency known as phenylketonuria (see Section 3.4). The excess of phenylalanine accumulated is converted into phenylpyruvic acid, by the path indicated by arrow e; this compound is then found in the urine of phenylketonurics and the reaction can be used for diagnosis of the disease. Tyrosine is converted into various other compounds by other enzymes. Blockage in the pathway that leads to formation of melanin, which is responsible for the dark color of skin, hair, etc., results in albinism. The block may occur at any step (only two are indicated, c and d).

but the overall result is phenotypically the same—albinism—if we limit ourselves to inspection of skin color. (It may be mentioned in passing that types of localized albinism are also known, indicating that some genes may affect only very specific parts of the body, such as particular tissues or organs.) If there were only one enzymatic step controlling the development of the pigment, the deficient allele being recessive, we would expect that a marriage of two albinos would give rise *only* to albino progeny. This is, in fact, the last cross in Table 3.1. But there are at least two different mutations that can lead to albinism, and something is known about them biochemically. Let us call A the dominant allele, forming the normal phenotype, at each albino locus, and a the recessive, deficient allele. Let us use the subscripts 1 and 2 to differentiate the two different genes (loci). These

TABLE 3.1
Six types of crosses for autosomal genes.

Parental genotypes	Progeny genotype			Progeny phenotype	
	gal+ gal+	gal+ gal−	gal− gal−	gal+	gal−
gal+ gal+ × gal+ gal+	all	none	none	all	—
gal+ gal+ × gal+ gal−	½	½	none	all	—
gal+ gal+ × gal− gal−	none	all	none	all	—
gal+ gal− × gal+ gal−	¼	½	¼	¾	¼
gal+ gal− × gal− gal−	none	½	½	½	½
gal− gal− × gal− gal−	none	none	all	—	all

genes may be on different portions of the same chromosome or on different chromosomes.

In this situation with two independent loci controlling the production of pigment, we may have three types of cross between albino parents, giving rise to the progeny listed in Table 3.2. The first two cases are straightforward. The third may need a slightly longer explanation; see Table 3.3, where the full genotypic composition of each parent is given. This table shows that the children are heterozygous for both loci. At each locus (1 and 2), the one normal allele makes enough of the enzyme to prevent the children from being albino.

Mutations at different loci that have a similar phenotypic effect are sometimes called *mimics*. For albinism, two mimic genes are known. There are many conditions that are at least partly genetic in origin for which many mimic genes are known. One of them is blindness; another is deaf-mutism; a third is mental deficiency. All of these may be due to mutations (dominant or recessive), but also can be due to traumas and other causes. When they are genetic, they are found to recur in families according to a dominant or to a recessive pattern of inheritance. It is not surprising that for diseases, such as blindness,

TABLE 3.2
An example of mimic genes (albinism).

Albino parents	Progeny
$a_1a_1 \times a_1a_1$ ⟶	all albino (a_1a_1)
$a_2a_2 \times a_2a_2$ ⟶	all albino (a_2a_2)
$a_1a_1 \times a_2a_2$ ⟶	all normal

TABLE 3.3
Fuller explanation of the last cross shown in Table 3.2.

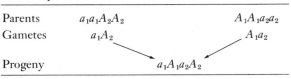

Parents	$a_1a_1A_2A_2$	$A_1A_1a_2a_2$
Gametes	a_1A_2	A_1a_2
Progeny		$a_1A_1a_2A_2$

deaf-mutism, and mental deficiency, that affect the functions of complex organs, there can be a very large number of mimic genes (from dozens to hundreds). The building of such complex organs demands the cooperation of many genes, and a defect in one gene may be enough to make the whole machinery nonfunctional. Because of the existence of many mimic genes, it may happen that a deaf person marrying another deaf person from an unrelated family has fully normal children, even though both parents are homozygous for a single recessive gene.

3.4 Independent Inheritance of Genes Located on Different Chromosomes

Hemoglobin is one of the many proteins in our body, but its great importance is shown by the large amount present (14 grams in every 100 ml of blood). It serves as a carrier of oxygen from the lungs to the tissues. Its molecular structure is well known. Each molecule is made of four subunits, or more precisely, of two pairs of subunits. The two types of subunit, each of which is represented twice, are chains called alpha and beta, which are, respectively, 141 and 146 amino acids long.

Several hundreds of different mutations in the two chains of hemoglobin are known today; most are rare and do not cause serious diseases. Figure 3.3 indicates some of the known mutations. One of these mutations is a very famous one, leading to what is called hemoglobin S (where S stands for sickle-cell anemia). It is known that hemoglobin S differs from hemoglobin A, which is the normal type, by a change in the amino acid at position 6 in the beta chain. This amino acid is valine in hemoglobin S and glutamic acid in normal hemoglobin (A). Since these two amino acids have different electric charges, in certain conditions they can be easily differentiated from each other electrophoreti-

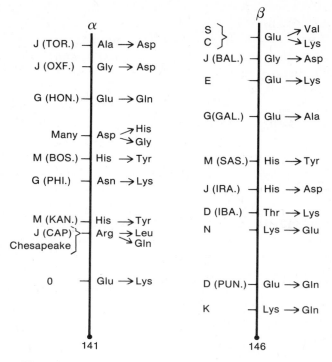

Figure 3.3

A sample of hemoglobin mutations, detected electrophoretically. At the left of each chain is the name (or one of the names) given to each mutant; at the right, the amino acid substitution involved. The abbreviations for amino acids are as follows: Ala (Alanine); Arg (Arginine); Asn (Asparagine); Asp (Aspartic acid); Cys (Cysteine); Gln (Glutamine); Glu (Glutamic acid); Gly (Glycine); His (Histidine); Ile (Isoleucine); Leu (Leucine); Lys (Lysine); Met (Methionine); Phe (Phenylalanine); Pro (Proline); Ser (Serine); Thr (Threonine); Try (Tryptophan); Tyr (Tyrosine); Val (Valine).

cally. The S gene is particularly frequent in some tropical areas of the world, Africa, and to a lesser extent India, but is also found in Greece and other parts of the Mediterranean. Because of their African origin, many black Americans have this gene. About 10% of all black Americans are heterozygous for this gene, and one in 400 is homozygous. Since the homozygote suffers from a fairly severe anemia, which drastically cuts his life expectancy, it is surprising that the

50   gene can reach such high frequencies. We shall see in more detail later that the gene attains this high frequency because the heterozygote has a capacity for survival higher than *both* homozygotes when malaria is present. At this moment we are interested in studying the S gene as a genetic marker, that is, an allele of the normal hemoglobin type, so that we can follow its fate easily in crosses.

We now consider another gene that affects some properties of the red blood cells, the Rh gene. Later (Chapter 5) we shall see its medical importance. Here it is enough to say that the r allele is recessive and gives rise to the phenotype called Rh−; all other alleles are Rh+. The Rh property is revealed in the red blood cells, being due to substances present on the surface of the cell, and has no known connection with hemoglobin. The gene controlling hemoglobin and the gene controlling Rh probably have nothing to do with each other, physiologically or genetically. We know they are not on the X chromosome; otherwise, they would show the typical sex linkage and pattern of inheritance that we saw in Chapter 2. Thus the two genes could be on any of the other 22 pairs, and the odds are roughly 21 to 1 that they are on different chromosomes.

We want to see the expectations for the offspring of crosses between individuals who differ with respect to two genes located on different chromosomes. Let us imagine a cross between two individuals:

Male $Hb^S Hb^A Rr$ × Female $Hb^A Hb^A rr$.

The father is heterozygous for both hemoglobin S and the Rh factor; the mother is homozygous for normal hemoglobin and is an Rh− individual.

What children do we expect from such a cross? We must follow the usual procedure, that is, list and count the gametes formed by each of the two parents, then combine the gametes in all possible ways and count the combinations. It is clear that the female can form only one type of gamete, $Hb^A r$, because she is a homozygote for each gene. But the formation of gametes in the male is a little more complicated. To understand fully the rules of gamete formation, we should always remember that (1) each gamete contains *only one member of each chromosome pair*, chosen at random, and hence there is a probability of 1/2 that a gamete contains a given member of a pair. Thus half the gametes will contain one of the two alleles present in the heterozygote and the other half will contain the other. But we should also remember that (2) *each gamete contains normally one member of all pairs of chromosomes*.

In Fig. 3.4 we represent what happens in the formation of the gametes 51 of the male parent in this cross. Figure 3.4, for reasons of simplicity, does not represent all 23 pairs of chromosomes, but just those two pairs that are marked by the two genes we are studying. In case A in Fig. 3.4, the random assortment of the members of each pair of chromosomes has put together in one sperm the gene Hb^S with the gene r, and in the other the gene Hb^A with the gene R; these two types of sperm will be in equal proportions. However, case A is not the only possible partition of chromosomes when a gamete is formed. There is no reason why the black chromosome of one pair must go with the black chromosome of another pair (here black and white refer to the representation of the chromosomes in the figure). Thus the distribution of the chromosomes can also be as shown in case B. Here two

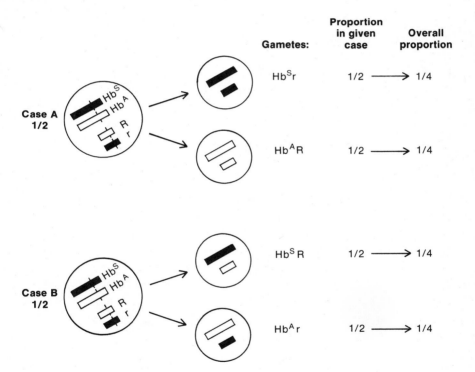

Figure 3.4
Case A: One possible mode of formation of gametes in the double heterozygote Hb^SHb^ARr. *The distinction between black and white chromosomes helps in visualizing the separation of chromosomes in gametes. Case B: The other possible mode of formation of gametes in the same heterozygote.*

52   types of sperm are formed, again in equal proportions, one of them $Hb^S R$ and the other $Hb^A r$. To determine the relative proportions of the two possible cases, A and B, one must remember that chromosome pairs do not influence each other when their members separate in the formation of gametes. Therefore, case A is as likely as case B, and if we pick one sperm at random it could be any one of the four types indicated in the figure, with an equal probability of 1/4.

As we have said, the eggs produced by the $Hb^A Hb^A rr$ female parent are always of the type $Hb^A r$. To find the genotypes of the individual offspring of this mating, all we have to do is put together the four types of sperm with the one type of egg ($Hb^A r$):

Zygotes formed

$Hb^S Hb^A rr$ $Hb^A Hb^A Rr$

$Hb^S Hb^A Rr$ $Hb^A Hb^A rr$

The chance of finding each of these four types is the same, because it depends directly on the proportions of the four types of sperm, which are all equal. Thus all four types of zygote are expected in equal proportions. This is an application of the *rule of independent assortment* of genes, which applies whenever two genes are located on different chromosomes.

3.5 Mendel's Laws

We may ask, at this point, what about Mendel's laws? Mendel basically expressed the results of his observation of crosses of garden peas in terms of numbers of the various types of plants he found in the progeny, and he interpreted the observed proportions of types in terms of simple probabilities, 1/2, 1/4, etc. He did not write specific "laws," but for didactical reasons it is convenient to summarize his main conclusions in the form of "laws" (two or more laws, depending on one's tastes).

1. In the cross between "pure lines" (homozygotes, as we call them today, say *AA* and *aa* for two alleles *A* and *a*) all progeny are genetically homogeneous, being made entirely of heterozýgotes *Aa*. They will all be similar between themselves; they will be intermediate in character between the two parents if neither parental trait shows dominance, or they will all be like one parent, and that parental form is then referred to as "dominant," and the other as "recessive." Usually one reserves capital letters for dominant alleles and lower case letters for recessive alleles.

2. In a cross between two heterozygotes (intercross), *Aa* × *Aa*, there   53
is a segregation (separation) into three classes: *AA, Aa, aa* in the proportions
1/4 : 1/2 : 1/4. We have seen that knowledge of chromosomes makes it easy
to predict that this should be the case. Mendel made the correct prediction
without this knowledge. In the case of dominance there are only two phenotypes, and they segregate in the proportion 3/4 dominant: 1/4 recessive.

3. In a cross between a heterozygous and a homozygous parent, say
Aa × *AA* (or *Aa* × *aa*), called a *backcross*, the two parental genotypes reappear
in the progeny in equal proportions.

4. Different genes (e.g., the pairs of alleles *A, a* and *B, b*) segregate independently; i.e., the combinations of genotypes obtained for each pair of
alleles considered by itself are simply obtained by multiplying the respec-
tive proportions. Thus, if in a given cross *AA* is expected in a proportion of
1/4 and *Bb* in a proportion of 1/2, the genotype *AABb* is expected in a proportion 1/4 × 1/2 = 1/8.

3.6 Autosomal Linkage and Assignment of Autosomal Genes to Specific Chromosomes

Early in this century it was found, in observations of experimental
crosses in plants and animals, that the rule of independent assortment
was sometimes violated. (At that time the rule was thought to apply
to *all* genes.) The chromosomal theory of inheritance explained these
exceptions. This theory predicted that genes located on the same
chromosome must show a different behavior from that of genes located
on different chromosomes. We have already seen what happens to
genes located on the X chromosomes (sex-linked genes). The situation
for genes located on the same autosome is similar to the one we have
already encountered with pairs of genes on the X chromosome, though
slightly more complex. If we can count the individuals whose genotypes indicate that crossing over has occurred and find what proportion of the total progeny of crosses these individuals represent, we
can estimate the percentage of recombination between two genes and
thus the distance between them on the chromosome. We should add
that if the two genes are on the same chromosome, but very far apart,
the gametes with crossing-over are equivalent in number to those with
no crossing-over. In other words, the percentage of recombination
is close to 50%. Thus, in practice, for genes that are far apart on the
same chromosome, all four possible types of gamete will have an equal

54 chance of being represented, just as we saw in Fig. 3.4 in the case of genes located on different chromosomes.

Therefore, when two genes are very far apart on a chromosome, it may be difficult to tell from observations on matings alone whether they do in fact belong to the same chromosome. There is, however, a way out. If a third gene is found which is between the two, it may be sufficiently close to each of them to show what is called *linkage* (that is, a percentage of recombination less than 50%) with both genes at either extreme. This linkage will show that the third gene is in the middle and that all three genes belong to the same chromosome (Fig. 3.5).

In theory, we should find that all genes fall into a number of groups called linkage groups, corresponding to the number of chromosome pairs.

Maps thus built are also called genetic maps. For organisms where experimental crosses are possible and where many mutations are known, such as the fruit fly *Drosophila*, corn, and mice, all mutants fall into linkage groups, each corresponding to a chromosome. For these organisms genetic maps have been drawn with good accuracy. Distances are expressed as percentages of recombination.

Another problem remains to be solved, however, and that is the assignment of specific genes or linkage groups to specific chromo-

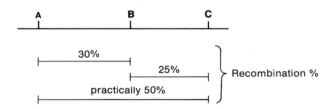

Figure 3.5
Genes A and C may seem not to be linked, because they are far apart on the chromosome and behave as if they were on different chromosomes. If a gene B is found, which is between A and C, it may show linkage (that is, recombination less than 50%) with both of them. This tells us that A, B, and C are all on the same chromosome, with B in the middle. Recombination percentages are imperfect measurements of distances between genes; they are not expected to add exactly. The distance (in percent recombination) between A and C is usually less than the sum of the distances between A and B and between B and C, and it cannot usually go above 50%.

somes. Some work has been done on man in this direction, and results have been obtained with a variety of methods. Now that staining techniques permit the accurate study of each chromosome, it has been found that some members of some chromosome pairs, although very similar, are not perfectly identical and can be distinguished. When such chromosomal peculiarities, usually small constrictions, are found in some individuals, they can then be studied in families in the same way that other genes, detectable in the general phenotype of the individual, can be studied. If two different alleles of a gene show linkage with identifiable morphological differences between members of a chromosome pair, it is possible to assign that gene to that chromosome.

Another method of assigning genes to chromosomes is that of using families in which chromosome aberrations are present, and studying them with respect to all other known genes. Chromosome aberrations determine specific alterations in the expectation of progeny types from crosses. If the chromosomes involved in the aberration contain some genes for which different alleles exist in the family in which the aberration is found, then it may be possible to locate some of the genes on specific chromosomes.

A third method, introduced more recently, can be used for all those characters, mostly of a biochemical nature, which can be studied in cells cultivated in test tubes. In this approach, cells, which may be of widely different animal origin, such as man and mouse, and which also show biochemically detectable differences, are fused together. The cells resulting from the fusion of one 46-chromosome human cell and one 40-chromosome mouse cell should have 86 chromosomes. Such cells tend, however, rapidly to lose many of these chromosomes, and one can isolate lines of cells that have lost specific chromosomes. By studying the biochemical properties of these exceptional lines and the chromosome losses they have undergone, one can often assign genes determining specific biochemical properties under study to the chromosomes that have been lost in these cell lines.

These and other methods have already permitted the assignment of some genes to specific chromosomes (Fig. 3.6). We can anticipate that this kind of research will produce many more results in the coming years. It might be worth mentioning some potential applications of such linkage studies. Suppose we know that an individual is affected by Huntington's chorea, a nervous disorder, which leads to the loss of movement coordination somewhat late in life. The disease is dominant and rare; individuals affected are heterozygotes and may marry

56

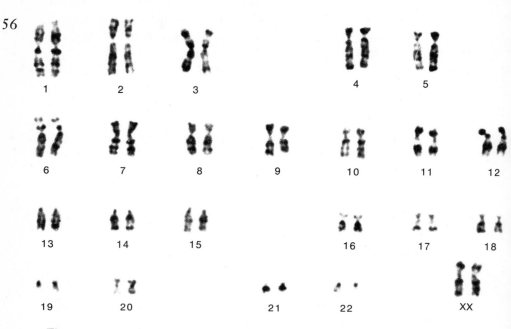

Figure 3.6
The chromosomes stained by G-banding shown at the beginning of the chapter are here paired and identified as to pair number. Linkage studies have permitted us to assign and occasionally locate with some precision individual genes to individual chromosomes. A list of such assignments can be found, for instance, in McKusick's catalog of Mendelian Inheritance in Man (see references).

normal individuals and have children. An individual whose parent (father or mother) turns out to be affected has a 50% chance of being affected himself (see Table 3.1) but may not know about his disease until late in life. If he wants to spare a child the agony to which he himself is exposed, he can only abstain from procreation, because at the moment we have no method of predicting whether a progeny will or will not develop the disease. But if we knew one gene to be closely linked to that for Huntington's chorea, and if we could study that gene in the fetus, we could tell during pregnancy with a good probability whether the progeny will or will not develop Huntington's chorea later in life and thus prevent its birth only if the disease is likely. For diseases for which preventive therapy is possible, linkage studies will offer the opportunity to improve our capacity to predict who is going to be affected. For this knowledge to be really useful in practice, however, we must wait for much more knowledge on linkage than is available so far. Linkage information has accumulated at a slow rate, but the rate is increasing all the time, and our maps should become much more satisfactory in relatively few years.

PROBLEMS

1. Phenylketonuria (PKU) is an autosomal recessive. If a child with this disease is born to two apparently normal parents, (a) what is the genotype of the parents? (b) what is the chance that another child born to the same two parents will have PKU? 25%

2. If individuals who have sickle-cell anemia, or those who are heterozygous for the gene (i.e., have the sickle-cell "trait" but not the disease) avoided marrying individuals who have anemia or who have the trait, would their offspring have sickle-cell anemia? No

3. If dark hair were inherited as a single dominant (it is only approximately so), could (a) husband and wife who are both dark haired have blonde children? (b) husband and wife who are both blonde have dark haired children? N.

4. The incapacity to taste a substance known as phenylthiocarbamide (PTC) is inherited as a recessive. If a taster, whose father is a non-taster, marries a non-taster, what will their progeny be like?

5. A blind man marries a blind woman. They both have blind relatives, and it looks very much as if both are homozygous recessives. What will their progeny be like?

6. A heterozygote for both the sickle-cell trait and for the Rh factor (*HbᴿHbᴬRr*, as in the example in the text) marries another double heterozygote like himself. What is the expected proportion, in the progeny, of sickle-cell anemia Rh negative individuals, sickle-cell anemia Rh+, non-anemic Rh−, and non-anemic Rh+? (This problem is more advanced than any given in the text. By noting what gametes are formed by each parent and in which proportion, and by forming all the possible combinations of the parental gametes, you should be able to come up with the right answer, which was first given by Mendel.)

Chapter four

The Hardy-Weinberg rule helps us understand the genetic composition of a population. We can simulate it by using two packs of cards, one for the male gametes and one for the female gametes. Assume a face card means a G allele and a non-face card a g allele. Taking one card from each pack we have: top right, a GG homozygote; bottom left, a Gg heterozygote; bottom right, a gg homozygote.

How Many Individuals of a Given Type?

Individuals forming a population differ from each other genetically. Some simple rules help us greatly in simplifying the description of this variation, and they allow some interesting predictions.

4.1 Dominant and Recessive

We are now fairly familiar with such terms as gene, allele, dominant, recessive, gamete, zygote (homo- and hetero-), phenotype, and genotype. There is some tendency to complain about an excess of terminology in genetics, but I cannot share this view. In fact, to understand the basic genetic phenomena, we need to know very few more than these ten terms. The difficulty is not so much in the number of terms, but rather in the fact that, as in other branches of science, we have to be careful to remember the exact definitions. Occasionally definitions are not as tight as one would like them to be. An example of a confusion that often arises is the confusion about the word *dominant*. One definition is that, in a heterozygote for two alleles, only one of the two

59

60   (the dominant allele) is expressed phenotypically; in this sense it dominates the other. A wrong conclusion which is often drawn, incidentally, is that dominant phenotypes should be frequent in populations, and that recessive ones should be rare because they are masked by dominants. But this is not necessarily true, and here we cite an example. A special form of dwarfism known as achondroplasia (sometimes referred to also as chondrodystrophy, produced by abnormal and contorted growth of the long bones) is due to a dominant gene, but it is very rare in the population, and affected people have few chances of reproducing. Thus most of them do not have descendants, and a very large majority of the population does not carry this gene even though it is dominant. What we usually want to know are the proportions in the population for *three* types of individuals: the homozygotes for the normal, recessive gene (non-dwarf ones); the heterozygotes for the dwarfing gene, whom we know are dwarfs; and the homozygotes for this same gene. These homozygotes, however, must be extremely rare; they can arise only in matings between heterozygotes, which are very rare themselves. Only a few putative homozygotes have been observed; they died very early with extreme signs of the disease. Thus, it might be argued that this type of dwarfing is not dominant at all, in the sense that the heterozygote is sort of intermediate in phenotype between the two homozygotes. Still, it is useful to retain the use of the word dominant in this case, to distinguish it from truly recessive disease. For instance, there are many other types of dwarfism; in particular, those due to deficiency of growth hormone (produced by the gland known as the pituitary) are fully recessive. Here, heterozygotes cannot be distinguished at all from normal homozygotes except for the fact that, when two of them marry, they have dwarf children (with a probability of ¼). A question that arises is the following: can we estimate how many heterozygotes for recessive dwarfism there are in the population? This and other questions can be solved by the methods which we consider in this chapter.

4.2 Frequency and Probability

We want to know the proportions of the three genotypes that are possible for two alleles at a locus in a given population. Codominant alleles are those which manifest themselves even if present in a single dose.

Here there is no problem: we can recognize the three genotypes, and therefore we can just count them. Take, as an example, the Rh gene. In Caucasian populations we know already that the Rh− allele is recessive, so that Rh+ Rh− individuals cannot be distinguished from Rh+ Rh+ (at least by the simplest technique). The Rh+ allele can, however, be further subdivided. For instance, appropriate reagents test for the presence of suballeles of Rh+ called R_1 and R_2. To make our task easier, let us consider Japanese instead of Caucasians, since Orientals do not have the Rh− allele. The R_1 and R_2 alleles are found also among Caucasians; in fact there are more than two dozen "alleles" known altogether. The structure of the Rh "gene" is probably a complex one, with many closely linked genes, but we want to simplify our task here by limiting our consideration to two "alleles" at a time. Thus one can (Table 4.1) score Japanese into the three categories: R_1R_1, R_1R_2, R_2R_2. Family studies have shown that the first and last genotype are made by homozygotes for allele R_1 and R_2 respectively, and the R_1R_2 type is heterozygote. If we score 100 Japanese and find 36 of them to be R_1R_1, 48 R_1R_2, and 16 R_2R_2, we know that the proportions of the three types are $36/100 = 0.36\ R_1R_1$, $48/100 = 0.48\ R_1R_2$, and $16/100 = 0.16\ R_2R_2$. These proportions are also called the (relative) *frequencies* of the three genotypes. A relative frequency is the ratio between the number of individuals having a given trait and the total number of individuals observed. Thus, if there are 24 students wearing glasses in a class of 50, the frequency of glasses-wearing students in that class

TABLE 4.1
Testing for two suballeles of Rh+ (R_1 and R_2) in a Japanese population. The nature of the reaction is similar to that used for ABO blood groups, which will be described in Chapter 5. The symbols + and − refer to the result of the test. Rh negative individuals, if any are present in the population, would score negatively with both reagents.

Genotypes	Reaction with reagents	
	Anti-R_1	Anti-R_2
R_1R_1	+	−
R_1R_2	+	+
R_2R_2	−	+

62   is 24/50 = 0.48. A relative frequency cannot be less than zero or greater than one. In what follows, we will omit the adjective "relative," although sometimes confusion may arise because the word "frequency" is sometimes used also for the actual numbers observed, in which case, strictly speaking, it should be called "absolute frequency."

An important consideration is that an observed frequency is rarely equal to that which can be obtained in a second set ("sample") of individuals drawn from the same population. Thus, another 100 Japanese would rarely show exactly the same frequencies of the three genotypes. It may be argued that the Japanese population is not strictly homogeneous, for there are some slight differences between northern and southern Japanese, for instance. Even if this were not true, however, two independent samples of the same population very rarely give exactly the same values. The discrepancy is called sampling error. If we take ten cards from a pack, we may find that four are red and six are black; we know there should be five and five, if there were no sampling error. If we mix our cards and repeat the procedure, we may come up with six black and four red, or three black and seven red, and only sometimes will we come up with five and five.

The way to minimize the relative importance of the sampling error is to increase the sample size (the number of individuals in the sample). If we increase the sample size so as to include the whole population, we would have the correct values, but this is very rarely feasible. The correct values correspond to the *probabilities* (or "true" frequencies) of the three genotypes. We must be content, usually, with the estimates of probabilities that frequencies observed in samples can generate. In the next three sections, however, we will reason as if we knew or could estimate the true frequencies—that is, we will temporarily ignore sampling errors, and will return to these at the end of the chapter.

4.3 Gamete or Gene Frequencies

Our aim will be to compute the proportions of homozygotes and heterozygotes for these two alleles of the Rh gene that we expect to find in the next generation that is produced by our sample of Japanese, 36 of which are R_1R_1, 48 R_1R_2, and 16 R_2R_2. The rules to be followed for this prediction are exactly the same as those we follow to predict the outcome of a single cross, except that here we are studying a great number of crosses all at the same time. We will have to determine the frequency of each of the two types of allele in both the male and the female gametes, combine the male and female gametes at random, and count the genotypes that arise from these combinations and the phenotypes that we can distinguish. An understanding of the relevant rule (called, after its discoverers, the Hardy-Weinberg rule) is absolutely essential to a study of human populations.

Let us imagine an experiment in artificial insemination, a technique sometimes used by couples unable to have children, because of the sterility of the husband. The child is conceived by injecting sperm obtained from another individual into the vagina of the female. It is customary to prepare a pool of sperm from many donors (who usually remain anonymous), so that it is usually not possible to determine who the biological father of a particular child really is. In general, donors are chosen from among healthy male students who want to earn a little money. As far as we know, neither the R_1 nor the R_2 type has any effect on health, and therefore the probability that the donor is of a given one of the three genotypes we have just enumerated is determined simply by their proportions.

Recall the operations that we described in predicting the outcome of a cross. The first operation is finding how many genes of a given type are present in the gametes. The sample of 100 volunteers includes 36 R_1, 48 R_1R_2, and 16 R_2 people. We know that in the formation of gametes the process of reduction or meiosis has taken place, and this has separated the two members of each chromosome pair. Therefore, in sperm from R_1R_2 individuals we will find two types of sperm cells, in equal proportions: one containing the R_1 allele, the other containing R_2. Forty-eight individuals out of 100 were R_1R_2; that is, the relative frequency of R_1R_2 individuals is 48/100 = 0.48 (or 48%). Half of the sperm cells of these R_1R_2 individuals (0.48 × 0.5 = 0.24 or 24%) will be R_1 and the other half, 0.24, will be R_2. In the individuals of type R_1 only one type of sperm is formed, R_1. The relative frequency of those individuals is 0.36 (or 36%), and therefore 0.36 of the sperm that go into the pool will consist entirely of type R_1. This 0.36 can be added to the 0.24 proportion of R_1 sperm from the R_1R_2 individuals, to give a total of 0.36 + 0.24 = 0.60. Similarly, the 16 individuals out of 100 (16%) who are of type R_2 form only sperm of type R_2, and therefore 0.16 can be added to the 0.24 proportion of sperm of type R_2 formed by the heterozygotes (0.16 + 0.24 = 0.40). We now can form a balance sheet as shown in Table 4.2. In this balance sheet the relative frequencies are expressed as true fractions, namely, as values smaller

TABLE 4.2
Computation of gene or gamete frequencies from genotype frequencies.

Genotype	Frequency of genotype	Gametes produced R_1	R_2
R_1	0.36	all (0.36)	none
R_1R_2	0.48	½ (0.24)	½ (0.24)
R_2	0.16	none	all (0.16)
Sums	1.00	0.60 (60%)	0.40 (40%)

than one, but it is customary to transform them into percentages by multiplying by 100. Thus we could say that R_1 gametes have a frequency of 60% and R_2 gametes have a frequency of 40%. In all numerical operations on relative frequencies which we will carry out later, if percentages are used, it is important to remember to convert percentages back to the fractional form of relative frequency. For instance, if one has a relative frequency expressed as the percentage 97%, one should remember that this means 97/100 or 0.97. In this way, one will avoid a type of computation error frequently made by beginners.

The two frequencies that we have just obtained, 60% and 40%, could also be called the frequencies of genes R_1 and R_2 in the gametes. Very often they are called simply *gene frequencies*. Allele frequencies in the gametes and gene frequencies in the population formed by the individuals who produce gametes are usually identical. It should now be clear that the gene frequency, i.e., the frequency of a given allele of a gene, can be obtained by a very simple arithmetic operation: summing the frequency of the homozygotes for that allele and half the frequency of the heterozygotes for that allele.

4.4 Proportion of Homozygotes and Heterozygotes in the Next Generation

When we consider a gene that is on a non-sex chromosome (i.e., an autosomal gene), we expect the same distribution of types among males as among females. In the following, we shall ignore sampling differences and imagine that a sample of males and a sample of females give exactly the same numbers of the three genotypes. This slight simplification does not alter the argument in any essential way.

We have so far computed the gamete frequencies, or gene frequencies, in the males, but the frequencies of the three genotypes should be the same in females as in males. There may be small differences, but we can usually ignore them. We can thus use, for the gene frequencies of the females, i.e., the frequencies in the eggs, the same values that we obtained among the males: 60% for R_1 and 40% for R_2. What we have to do next is mate the male and female gametes at random, an operation which may seem somewhat artificial. It might be less artificial if we could subject the female gametes, the eggs, to a pooling operation similar to that which is done with sperm in artificial insemination. In reality, in ordinary human fertilizations one male and one female mate. But where there is no choice of husbands by wives or vice versa for the gene we are studying, we can use the term *random mating*, and it can be shown that the random mating of pairs of adults will give the same results as making a pool of gametes and letting them mate at random. By "at random" we mean that every sperm has the same probability of fertilizing any egg existing in the pool. The random mixture and mating of gametes is represented graphically in Fig. 4.1. In this figure we have a number of pairs, each consisting of one sperm and one egg. Sperm of type R_1 and eggs of type R_1 are represented in black, and gametes of type R_2 are represented in white.

We can compute more easily and see more directly the effects of this random union of gametes by looking at Fig. 4.2, in which the various possible pairs have been ordered and represented in frequencies equal to the true ones. We recognize that four possible types of pairs are formed: sperm R_1 with egg R_1, sperm R_1 with egg R_2, sperm R_2 with egg R_1, and sperm R_2 with egg R_2. The proportions of the four types are represented graphically by the area they occupy. A very simple geometric consideration, which can be understood by looking at Fig. 4.3, shows that we can expect the mating of sperm R_1 with egg R_1 to have a frequency equal to the product of the relative frequencies of the two gametes, that is, $0.6 \times 0.6 = 0.36$, and therefore 36% of all zygotes will be of type R_1R_1. On the other hand, the other type of homozygote, R_2R_2, can be expected to constitute $0.4 \times 0.4 = 0.16$, or 16% of the progeny. All other matings form heterozygotes, and these heterozygotes might be divided into two classes: those in which it is the sperm that is R_1 and the egg R_2, and the complementary type, where the sperm is R_2 and the egg R_1. However, in general, since we are not interested in distinguishing these two types of heterozygotes,

Figure 4.1
Random fertilization of egg by sperm. Each sperm and each egg carries 23 chromosomes, but only one gene R is being considered, and the head of the sperm or the nucleus of the egg is shown as black if gene R is of type R₁, white if gene R is of type R₂. In both sperm and eggs the ratio of R₁ to R₂ types is taken to be 3:2 (that is, 60% of type R₁ and 40% of type R₂). Only one sperm is shown fertilizing each egg, because all possible latecomers are excluded from fertilization. After the meeting of a sperm with an egg, the sperm penetrates the egg (penetration is not represented here or in Fig. 4.2) and the nucleus of the sperm fuses with that of the egg, producing a cell with 46 chromosomes: the zygote. Relative sizes of sperm and eggs are altered for simplicity.

we pool them. The first type has a frequency of $0.6 \times 0.4 = 0.24$, the second has a frequency of $0.4 \times 0.6 = 0.24$, and the sum of both is $0.24 + 0.24 = 0.48$.

Summarizing, we have $0.36\ R_1R_1$ individuals, $0.48\ R_1R_2$ individuals, and $0.16\ R_2R_2$ individuals (see Fig. 4.3). This is exactly the same as the frequencies with which we started. If we repeated the operation for one or more successive generations, clearly we would always end up with the same gene and genotype frequencies at every generation.

4.5 The Hardy-Weinberg Rule and Some Applications

What we have just seen is a numerical application—to a specific example—of a more general rule that goes under the name of Hardy-Weinberg. The results we saw in the earlier section can be easily generalized. Let us use the letter p to represent the gene frequency of R_1, and q to represent the gene frequency of R_2. If these two are the only alleles that we can distinguish in the population, then $p + q = 1$. Thus

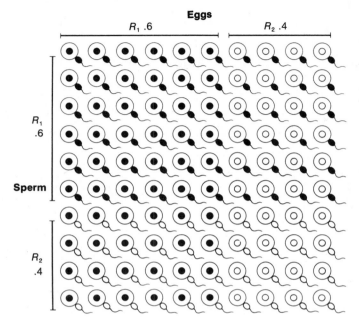

Figure 4.2
One hundred zygotes obtained from a random fertilization, as in Fig. 4.1, with 60% of gametes (sperm or egg) of type R_1 and 40% of type R_2. They are ordered with all R_1-carrying eggs at the left, R_2 eggs at the right, R_1 sperm on top, and R_2 sperm on the bottom. The expected proportions of zygotes R_1R_1, R_1R_2, and R_2R_2 are also given in geometric form in Fig. 4.3. Homozygotes R_1R_1 and R_2R_2, respectively, are in the upper left and the lower right quadrants; the rest of the square is occupied by heterozygotes R_1R_2.

the frequency with which the R_1R_1 type is to be expected in the next generation is simply $p \times p = p^2$. The expected frequency of R_1R_2 is $2pq$, and that of the R_2 homozygote is q^2. This is called the Hardy-Weinberg rule and is of central importance to all the genetics of diploid organisms (Table 4. 3). We now want to make some comments to demonstrate, to some extent, the importance of this theorem and some of its applications.

1. *Constancy of gene frequencies.* If we count the gene frequencies of R_1 and R_2 in other populations of Japanese males and females, we will find that it remains the same from generation to generation. This is, of course, not surprising. Imagine a bag of beans, some of which are white and some of which

	Eggs	
	$R_1 = .6$	$R_2 = .4$
$R_1 = .6$	.6 x .6 = .36 or 36% $R_1 R_1$ Homozygotes	.6 x .4 = .24 or 24% $R_1 R_2$ Heterozygotes
$R_2 = .4$	.6 x .4 = .24 or 24% $R_1 R_2$ Heterozygotes	.4 x .4 = .16 or 16% $R_2 R_2$ Homozygotes

Sperm (label at left, between the two rows)

Figure 4.3

The hundred zygotes of Fig. 4.2 are represented as forming a square, subdivided into four parts. The proportions of eggs and sperm of types R₁ and R₂ are indicated on the sides, and the four areas in the square represent the four types of zygotes, two of which, representing the heterozygotes (top right and bottom left), can be pooled together.

are black in specified proportions. If all the beans are transferred to another bag with none being lost, the proportion of white to black will obviously remain the same. The transfer from one bag to the other corresponds, in the case of genes, to the shift from one generation to the next. Thus, if the analogy holds, gene frequencies do not change unless some outside forces intervene. The forces which change gene frequencies are evolutionary ones, and we will consider them in Chapter 5. In their absence, gene frequency remains constant.

2. *The Hardy-Weinberg equilibrium is reached in one generation.* If we determine the gene frequency in the new generation that we have obtained, we find that it is equal to that of the former generation. In fact, the frequencies of the three genotypes that we have obtained are identical to those with which we started. The frequencies of genotypes with which we started were chosen in order to make the calculations as simple as possible. But suppose we started with different initial frequencies of genotypes, with the restriction that the gene frequency must be the same. Imagine, for instance, that both in males

and females the frequency of R_1 in the initial sample is 50%, that of R_1R_2 is 20%, and that of R_2 is 30%. This distribution of genotypes (50, 20, 30) is quite different from the one we started and ended with before (36, 48, 16). It happens, however, that these values represent the same gene frequencies, $p = 0.6$ and $q = 0.4$, as can easily be verified. The value of p, the gene frequency of R_1, is $0.50 + \frac{1}{2}(0.20) = 0.6$. Similarly, $q = 0.30 + \frac{1}{2}(0.20) = 0.4$. Applying the Hardy-Weinberg rule we know that in the next generation we will find the proportions given by the expression $p^2 + 2pq + q^2$, namely 36%, 48%, and 16%. Moreover, these proportions will remain the same in all successive generations. Therefore, applying the Hardy-Weinberg rule, the distribution of genotypes reaches in one generation an equilibrium among the three genotypes which remains constant in all successive generations (the Hardy-Weinberg equilibrium is that summarized in Table 4.3). Thus, not only do the gene frequencies remain constant if there are no evolutionary forces at play (as we said before), but if there is random mating, the genotype frequencies remain constant.

3. *More than two alleles.* The Hardy-Weinberg rule easily extends to more than two alleles, as we shall see in Chapter 5.

4. *Recessives.* One of the most common applications of the Hardy-Weinberg rule is to predict the frequency of homozygotes and heterozygotes when there is complete dominance, and we cannot, therefore, separate the dominant phenotype into the two genotypes that contribute to it. The example we shall consider first is that of a *rare recessive*, such as phenylketonuria (PKU), a defect due to the lack of, or insufficient activity of, an enzyme that transforms phenylalanine into tyrosine (see Fig. 3.2). Because of the accumulation of phenylalanine that thus results, and the relative lack of tyrosine, phenylketonurics differ from normal people in several ways, but the major difference is often a severe mental deficiency. This deficiency can be avoided only if during the critical period of development—the first months of life—their diet is regulated in such a way that their intake of phenylalanine is much smaller than that of normal individuals. The frequency of individuals who are born with phenylketonuria varies somewhat from one population to another, but is approximately 1/15,000 among U.S. whites.

TABLE 4.3
The Hardy-Weinberg equilibrium.

Genotype	R_1R_1	R_1R_2	R_2R_2
Phenotype	R_1	R_1R_2	R_2
Expected proportion	p^2	$2pq$	q^2

p is the frequency of gene R_1, and is equal to the frequency of genotype R_1 plus half that of heterozygotes; q is the frequency of gene R_2, and is equal to $1 - p$, and also to the frequency of genotype R_2 plus half the frequency of the heterozygote.

70   The study of the inheritance of phenylketonuria has shown that a recessive gene is involved. Phenylketonurics usually do not reproduce. Clearly, therefore, the disease can be found only among the progeny of two individuals who are *both* heterozygous for the gene. Because the gene is recessive, these two individuals will be normal. There are tests, which are not absolutely reliable, to tell whether individuals are heterozygous, but they must be carried out in specialized laboratories. According to Mendel's laws, the probability that two heterozygotes will have a phenylketonuric child is 1 in 4. We ask the question, "How many heterozygous individuals are in the population?" The answer can come, via the Hardy-Weinberg rule, directly from the knowledge of the frequency of PKU homozygotes at birth. This, we said, is 1/15,000 (=0.00007), a value which can be equated with the frequency of the homozygotes q^2, if q is the frequency of the gene for phenylketonuria. The value of q can easily be obtained as the square root of 0.00007, that is, about 0.008. The frequency of heterozygotes is given by the Hardy-Weinberg rule as $2pq$, and we can calculate $2pq = 2(1 - 0.008)(0.008) = 0.016$ approximately, or 1.6%. Thus, one individual out of 60 is heterozygous for PKU. This may seem to be a high value, and it is, but it should be remembered that these individuals are phenotypically normal, at least in the sense that they have no known phenotypic handicaps.

Table 4.4 gives, for a few selected gene frequencies, some values of genotypic proportions, and ratios of heterozygotes to dominant phenotypes when an allele is dominant.

The Hardy-Weinberg rule is one of the strongest laws in genetics and it is difficult to find deviations from it. If such deviations are found, they may have a variety of meanings, all of which may be interesting.

5. *Equilibrium for sex-linked genes.* Genes located on the X chromosome will behave in a somewhat peculiar way from the point of view of the Hardy-Weinberg rule. Females follow the rule, since they have two X chromosomes: but males have only one X chromosome, which they receive from their mother. Consequently the frequency of a given allele in males will be equal to the gene frequency of that allele in their mothers. This fact makes computations simple, but it introduces the complication that males and females have different proportions of an X-linked character. For example, in a population of black Americans it is usually found that 10% ($q = 0.1$) is the frequency among males of the gene $G-$ (or g), that is, the lack of G6PD which we have already studied, and that the frequency of $G+$ (or G) among males is 90% ($p = 0.9$). The frequencies found in the males are equal to the frequencies among the gametes of the mothers, and therefore can be used as values of p and q to compute the Hardy-Weinberg equilibrium expected in the females. Thus, if we want to know, for instance, how many women are homozygous *gg* and therefore G6PD-deficient, we should compute it (from the Hardy-Weinberg rule) as q^2. We find it to be equal to $0.1^2 = 0.01$, and therefore we can predict that the proportion of G6PD-deficient women in that population will be 1%. Far fewer women are affected than men, as is true of any recessive sex-linked genes.

6. In all the computations above we have ignored *sampling errors*. When examining observed data, however, we cannot ignore them; to them we dedicate the next section.

4.6 Sampling Errors

There is a simple way to form a firsthand idea of the magnitude of sampling errors: simulate random sampling with some procedure for producing random events—for instance, coin tossing, dice throwing, game cards drawing, computers, etc. We will use game cards as an example, and first take out all the aces, thus leaving 48 cards. If we take one card out of this pack and score it according to whether it is a face card or not, we imitate an event with a probability of 1/4. There are in fact 12 face cards in a pack of 48, and 12/48 = 1/4. Calling a face card an *A* allele, and a non-face card an *a* allele, we simulate gene frequencies $p = 0.25$ (or 1/4) and $q = 0.75$ (or 3/4) respectively for *A* and *a*. A diploid individual can be simulated by taking a card from one pack (say, the male gametes) and another from another similarly prepared pack (the female gametes). Two faces is an *AA* individual, one

TABLE 4.4
Hardy-Weinberg equilibrium: genotypes and phenotypes expected for some gene frequencies.

Gene frequencies of		Genotype frequencies of (%)			Approximate percentage of heterozygotes *Aa* among all *A* types (*AA* and *Aa*) if *A* is dominant
A	*a*	*AA*	*Aa*	*aa*	
0.999	0.001	99.8%	0.1998%	0.0001%	0.2%
0.99	0.01	98%	1.98%	0.01%	2%
0.9	0.1	81%	18%	1%	18.2%
0.8	0.2	64%	32%	4%	33%
0.7	0.3	49%	42%	9%	46%
0.6	0.4	36%	48%	16%	57%
0.5	0.5	25%	50%	25%	66%
0.4	0.6	16%	48%	36%	75%
0.3	0.7	9%	42%	49%	82%
0.2	0.8	4%	32%	64%	89%
0.1	0.9	1%	18%	81%	94.7%

TABLE 4.5
Results of simulating Hardy-Weinberg equilibrium using
two packs of game cards, with $p = \frac{1}{4}$, $q = \frac{3}{4}$, and 100
individuals.

	AA	Aa	aa	Total
Experiment 1	10	34	56	100
2	6	33	61	100
3	4	43	53	100
4	9	36	55	100

face and one non-face a heterozygote *Aa*, and two non-faces an *aa*.
By putting the cards back into the pack to which they belong, mixing
the packs, and repeating the operation, we get another individual, and
so on. Although in practice it may prove rather tedious, this procedure
shows that it may take a very large sample to get close to the expected
proportions of $(1/4)^2 = 1/16\,AA$, $2(1/4)(3/4) = 6/16\,Aa$, $(3/4)^2 = 9/16\,aa$.
To speed up the exercise by using teamwork, we could have one person keeping a deck (say, the male gametes) and shuffling it, a second
person doing the same for the female deck, a third person picking a
male and a female gamete (a card from the male deck and a card from
the female deck), and a fourth person scoring the results. Clearly, it
is even faster to use a computer. The main results to observe in this
exercise are the following: (1) in a very small sample, say ten individuals, the deviations are large; (2) if we increase the size of the
sample, the deviations from the expectation get smaller. Gene frequencies other than $p = 1/4$ and $q = 3/4$ can be imitated by using other
rules, e.g., $p = q = 1/2$ by scoring for red or black cards. Had we left
aces in the pack, the probability of a face card would have been $3/13 =$
23%, approximately, and this is then the p gene frequency we would
have simulated, etc. Results from several such experiments, for $p =$
$1/4$ and $q = 3/4$, each on 100 individuals, are given by way of example
in Table 4.5. In this way we form an idea of the magnitude of random
sampling error, but we do not solve the following practical problem:
if we observe among N individuals certain numbers of *AA*, *Aa*, and
aa individuals, do these data reflect Hardy-Weinberg equilibrium?
There is a statistical method which can answer the problem, and its
use is illustrated in Table 4.6. We thus compute, from a set of observed
numbers, the agreement with the hypothesis of Hardy-Weinberg,
in the form of a statistical index called χ^2 (chi-square). If the computed

TABLE 4.6
The Chi-Square Test.

This is a test to determine whether a given deviation from the frequencies expected from the Hardy-Weinberg rule is compatible with random statistical fluctuation. Suppose in an observed sample the following numbers are found for the three genotypes of a gene having alleles A_1 and A_2:

AA	A_1A_2	A_2A_2	Total
55	200	245	$500 = N$

1. Compute the gene frequencies:

$$p \text{ (gene } A_1) = \frac{55}{500} + \frac{1}{2} \times \frac{200}{500} = 0.31$$

$$q \text{ (gene } A_2) = \frac{1}{2} \times \frac{200}{500} + \frac{245}{500} = 0.69$$

(Check: $p + q = 1$)

2. Compute the Hardy-Weinberg proportions of the genotypes:

A_1A_1	A_1A_2	A_2A_2
$p^2 = 0.31^2$ $= 0.0961$	$2pq = 2 \times 0.31 \times 0.69$ $= 0.4278$	$q^2 = 0.69^2$ $= 0.4761$

3. Multiply the Hardy-Weinberg proportions by the total number of individuals, $N = 500$ in this case:

A_1A_1	A_1A_2	A_2A_2
48.05	213.90	238.05

These are the expected numbers (E), based on the hypothesis that the Hardy-Weinberg rule will indeed hold true.

4. Carry out the operations indicated in the table below:

Genotype	Number observed (O)	Number expected (E)	$O - E$	$(O - E)^2$	$(O - E)^2/E$
A_1A_1	55	48.0	+ 7.0	49.0	1.02
A_1A_2	200	213.9	−13.9	193.21	0.90
A_2A_2	245	238.1	+ 6.9	47.61	0.20
Sum	500	500	0		$2.12 = \chi^2$

The value obtained, χ^2, corresponds to the formula

$$\sum \frac{(O - E)^2}{E},$$

where the sum is extended to all genotypes, and gives a measure of disagreement between the hypothesis (that the Hardy-Weinberg rule holds) and the data. If it is higher than 3.8 (for three genotypes), the hypothesis is usually rejected. This conclusion implies a level of confidence which can be stated, numerically, as having a probability (P) less than 5%, meaning that one can get an agreement as bad or worse by pure chance, if the hypothesis is true, less than 5% of the time. Here χ^2 is less than 3.8, so the hypothesis is accepted; i.e., the deviation observed can be due to random statistical fluctuations and thus the Hardy-Weinberg rule holds.

74   value of χ^2 is less than 3.8, we accept the hypothesis of Hardy-Weinberg equilibrium as being correct; if the value of χ^2 is greater, we reject the hypothesis. We should not delude ourselves, however, into thinking that this is the final answer to our problem. No statistical method or, for that matter, any method can ever tell us with certainty if a hypothesis is true or not. A χ^2 computed from one of the examples of Table 4.5, for which we know that the hypothesis of Hardy-Weinberg is correct, having generated the data accordingly, has a 5% chance of being greater than the threshold value of 3.8. Since we accept the hypothesis as being correct when χ^2 is less than 3.8 (see Table 4.6), we know that, of all occasions when the hypothesis is correct, we will wrongly reject the hypothesis no more than 5% of the time (on average). Thus we can expect that, one time out of 20, experiments like those with the pack of cards will be wrongly declared as not satisfying the hypothesis, but this is a relatively low risk and we have to accept a certain level, or risk, of error in any case.

PROBLEMS

1. The inability to taste PTC (a substance tasting bitter to 70% of Caucasians and having basically no taste for the other 30%) is due to a recessive autosomal gene. This statement needs some further qualifications, but here we will forget the complications. (a) What is the frequency of the recessive and the dominant genes? (b) What is the proportion of heterozygotes among tasters?

2. Most adult Caucasians are capable of digesting lactose (milk sugar). This is probably due to a single dominant gene. If 75% of adult Caucasians utilize lactose, what is the frequency of the relevant gene?

3. The recessive disease cystic fibrosis affects one out of every 3000 Caucasians born. What is the proportion of heterozygotes? (When a gene frequency q is small, you can take this proportion to be $2q$ instead of the correct $2pq$, since p is very nearly one.)

4. Hemophilia is an X-linked recessive disease. The frequency in males is about one out of every 20,000 born. What frequency would you expect among females, using the simple assumptions given in the text?

5. Test the validity of Hardy-Weinberg by computing χ^2 on one of the examples given in Table 4.5.

Chapter five

Blood groups are determined by agglutination reactions. Shown here is the reaction for ABO blood groups (subject is of A blood group). Above, the reagents, a microlance, and stir-sticks. Bottom left, two drops of blood and a drop of each of the two antisera; the drop of anti-B reagent and one drop of blood are being mixed. Bottom right, after both drops of blood have been mixed, each with its antiserum, and after about a minute has elapsed, one can notice agglutination of the blood red cells taking place in the drop at the left (which was mixed with anti-A reagent), and no change in the drop at the right (mixed with anti-B).

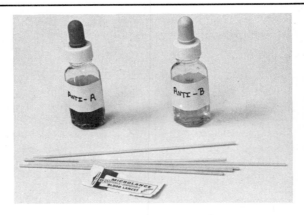

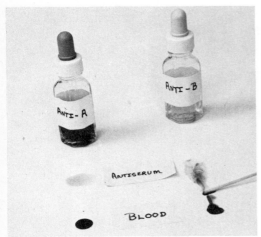

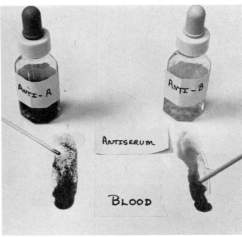

Immunity and Genetics

The body defends itself against attack from bacteria, viruses, etc., by the immune system. Immunological techniques are used for the study of "blood groups," some of which (ABO, Rh) have considerable medical significance.

5.1 Immunity, Antigens, and Antibodies

Vertebrates have developed a distinctive system of protecting themselves from the attack of foreign germs, viruses, and poisons. This system is based on the fact that, when in contact with foreign organisms or certain types of foreign substances, individuals develop antidotes which are usually specific against the particular intruder or damaging substance. These specific antidotes are called *antibodies*. They are all protein molecules, known as gamma globulins, and are usually found in the serum—the yellow liquid part of blood—remaining after blood coagulates and the red cells have collected in a red clot.

Antibodies are made by special cells, white blood cells and others, but the capacity to make antibodies does not develop immediately after birth. In the first months of life, during which an infant is unable

78   to make antibodies, it is protected by those it has received from the mother through the placenta. This protection is called *passive immunity*. Passive immunity may also be used for therapeutic purposes. For instance, after a potentially infectious wound has occurred, one way to protect people against tetanus infections is by injecting them with antibodies against the toxin that is produced by the tetanus bacillus. These antibodies have usually been made by horses that have been injected repeatedly with the tetanus toxin so as to detoxify it and yet leave unaltered its capacity to elicit antibodies.

After a few months an infant becomes capable of developing his own antibodies, and he can thereby develop *active immunity*. The production of antibodies is not entirely understood, but we know a number of important things about it. The average individual can make a great variety of antibodies. If an individual is attacked by the bacterium causing typhoid fever, he will produce antibodies of more than one type, all directed against specific substances contained in the bacterium. If the individual is not efficient in producing antibodies and if he is not helped by means of drugs specifically capable of destroying the bacterium or at least of slowing its growth, he may be unable to survive by his own forces. However, our immunological defenses readily dispose of most infectious diseases to which we are exposed.

Not all substances can induce formation of antibodies directed against them, but almost all viruses and bacteria or protozoa or other infective organisms contain substances against which the organism can produce antibodies, so that some degree of protection can be obtained. Foreign substances that evoke production of a specific antibody directed against them are called *antigens*. The reaction between an antigen and its antibody is highly specific, resulting in some sort of neutralization of the antigen's action. When an antigen is toxic, there is an obvious advantage in being able to neutralize its activity by a specific antibody. Because of the specificity of antibodies, and the great variety of antigens in existence, the organism must be able to produce an almost infinite variety of antibodies. The fact that the human organism is equipped with this complex system of protection—immunity—is the cause of many genetic complications and sequels.

5.2 ABO Blood Groups

At the beginning of the century it was discovered that one cannot transfuse blood from one person to another without some preliminary checks to ensure that the transfusion will be successful. This discovery led to the classification of all humans in four blood groups, now called ABO. Every individual belongs to one of the following blood groups: O, A, B, or AB. It is safe to transfuse blood from an individual of one group into another individual of the same group; but when blood is transfused from an individual of one group into an individual of another group, some strict rules must be followed. The penalty may be serious disease or even the death of the person who receives the foreign blood.

We know that the differences in ABO blood groups are due to chemical differences in substances present on the surface of red blood cells and on other cells in the body. These substances are fairly complex, but the important constituents, so far as blood groups are concerned, are some of their sugars. These sugars are attached to the molecules of the blood-group-determining substance by several varieties of enzymes. The enzymes are produced by genes, and differences in enzymes are due, as we have already said, to mutations that occurred a very long time ago in those genes.

Tests for ABO blood groups are very simple. We need two reagents called anti-A and anti-B respectively. They are usually antibodies, and we shall see later how they can be obtained. To test the blood group of an individual we need two drops of his blood. One of these drops is mixed with reagent anti-A and the other with reagent anti-B. After a few minutes either one or the other of two possibilities will be observed in each drop: (1) the blood will remain homogeneously red; this is a negative reaction. (2) Alternatively, the blood may collect in clumps, which upon shaking tend to break into smaller clumps only with difficulty. If we look at these two reactions under a microscope with fairly low magnification, we see that in a negative reaction the blood cells have remained separated from one another; in a positive reaction, on the other hand, they have come together in clumps. This clumping reaction is called *agglutination*. In a test of the blood of one individual, the reaction can be positive or negative with reagent anti-A and positive or negative with reagent anti-B. Altogether, four reactions are possible. They are illustrated in Fig. 5.1, which also indicates how individuals are classified in the four blood groups according to their reactions to anti-A and anti-B. Substances resembling blood group substances A and B are present in many other organisms, including bacteria. Thus it is not surprising that we manufacture antibodies against them.

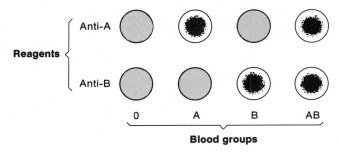

Figure 5.1
Assignment of an individual to one of the four ABO blood groups on the basis of the reactions of blood drops with anti-A and anti-B reagents. A dark circle indicates a negative agglutination reaction; the formation of red cell clumps indicates a positive agglutination reaction.

There is a general rule in immunity that one cannot develop antibodies against oneself. This rule is violated only very rarely, and then may give rise to diseases called "auto-immune." As regards ABO blood-group substances, the rule is that during life an individual develops antibodies against all the antigens that he does not possess at birth. From this point of view, O can be considered as absence of antigens of either the A or the B type. A and B individuals produce blood-group substance A and blood-group substance B respectively. It is thus possible for individuals of blood group B to produce antibodies against substance A, and they will usually do so. These antibodies will not harm the group B individual unless he or she is transfused with blood from a group A (or AB) individual, because in such an event, the red cells received would be agglutinated by the anti-A, which the group B individual contains. Thus an individual of blood group A cannot give his blood to an individual of blood group B and vice versa.

The antibodies that are found in the blood of individuals of the various blood groups are listed in Table 5.1, and from this list we can build the rules for permissible transfusions. A further principle to be kept in mind is that when one person gives blood to another, a small amount of the blood of the donor is mixed with the much larger amount of blood contained by the receiver. Suppose the blood donor is of group O. If the recipient is of blood group A, we might think his red cells would react with the anti-A from the donor and agglutination would occur, with subsequent damage to the recipient. If red blood

TABLE 5.1

Antibodies usually contained in the serum
for ABO blood groups.

If the red cells are	The serum contains antibodies
O	anti-A and anti-B
A	anti-B
B	anti-A
AB	none

cells of an individual agglutinate in his arteries, veins, or capillaries, the blood cannot circulate freely, and severe damage may result. However, because the amount of blood given is usually small, the antibodies contained in the serum of the donor are diluted by the blood of the receiver sufficiently so that they are usually not dangerous. Thus we can give blood from a type O individual to an individual of any group, including, of course, type O, because the antibodies contained in the serum will be diluted so as to be ineffective, and the donor's red cells do not contain substances that will react with the anti-A or anti-B contained in the blood of the receiver. Therefore, individuals of blood group O are called universal donors.

AB individuals, on the other hand, cannot give blood to any other group but AB, because their red cells would always find in the blood of other individuals antibodies that would agglutinate them. AB individuals may, as can be easily understood, receive blood from any other group, and therefore are called universal recipients.

The genetics of ABO can be described in simple terms. There are three alleles: A, B, and O (A can be further subdivided by finer tests, but we shall neglect this). With three alleles there are six possible genotypes, listed in the first column of Table 5.2. Because O is recessive to A and B, only four phenotypes are actually observed, and these correspond to the four blood groups that we have seen; they are given in the second column of Table 5.2. The last column of Table 5.2 contains the approximate proportions, in a Caucasian population, of the four blood-group phenotypes and the corresponding genotypes. It also indicates how these have been computed, by an extension of the Hardy-Weinberg rule, to three alleles. Take p as the given frequency of A, q as that of B, and r as that of O. A individuals can be AA or AO (because A is dominant over O). The frequency of the homozygote for an allele

TABLE 5.2
Proportion of ABO phenotypes and genotypes in a
Caucasian population. p is the gene frequency of A, q of B,
r of O. Here $p = 0.28$, $q = 0.06$, $r = 0.66$.

Genotype	Phenotype		Proportion in population
AA AO	A	45%	$\begin{cases} 7.8\%(p^2) \\ 37.0\%(2pr) \end{cases}$
BB BO	B	8%	$\begin{cases} 0.4\%(q^2) \\ 7.6\%(2qr) \end{cases}$
OO	O		$44\%(r^2)$
AB	AB		$3\%(2pq)$

is the square of the frequency of that allele, and this is p^2 for AA (q^2 for BB, r^2 for OO). The frequency of the heterozygote for two alleles is twice the product of the gene frequencies for those alleles, and is thus $2pr$ for AO. The total frequency of individuals of phenotype A is then $p^2 + 2pr$. Similarly one can compute the total frequency of B individuals. Since A and B are codominant, AB individuals are heterozygotes with frequency $2pq$. O individuals can only be homozygous for O, and their frequency is r^2.

An interesting application of blood groups, along with all other genetic markers (traits whose genetic transmission is well known and clear-cut), is in tests of disputed paternity. A man falsely accused of the paternity of a child has a good probability today of proving his innocence. Thus, a B child cannot be born to an O or A father, if the mother is O or A; the true father must be either B or AB. Naturally it is much more difficult to prove that a given individual *is* the father; any B or AB individual could be the father of that child. But as we increase the number of genetic markers, the circle of potential fathers can be considerably reduced.

5.3 Rh Groups

Many other blood group systems have been discovered since ABO. We shall mention here only one, Rh, because of its medical significance. We have already mentioned the Rh gene and used alleles R_1 and R_2, which we called Rh+, and allele r, which we called Rh−. For clinical purposes we can ignore the distinction between the many

Rh+ alleles in existence. It will be enough here to consider only an *R* allele, including all *R*'s mentioned as well as others we have not mentioned, and an *r* allele. These alleles, like the blood-group substances A and B, affect the production of substances found on the surface of red cells. These substances are chemically different from those of the ABO blood groups.

The Rh group came to the attention of clinicians in connection with childbirth. Women who are homozygous for the allele *r* (and are therefore Rh−) can have Rh+ progeny if the father is Rh+. When this happens, the child may be exposed to severe anemia while in the uterus, due to the destruction of his red cells and to other pathological consequences, which may even cause the death of the child before or shortly after birth. The reason is that the mother of the child has manufactured antibodies against the Rh+ antigens contained in the red cells of the fetus, and these antibodies have been passed to the fetus, whose red cells have, in part at least, been agglutinated and destroyed. If the loss of blood thus incurred by the fetus is very severe, the outcome is fatal. In cases of intermediate severity the result may be mental impairment or other deficiencies.

The possibility that the mother immunizes herself against the red cells of her fetus in utero may seem strange, since it is known that the circulation of fetal blood is separate from that of maternal blood. Nutrients and relatively large molecules (for instance, molecules of some antibodies) can go through the barrier between the two circulatory systems, but not red cells. There are, however, occasional leakages whereby the red cells of the fetus can enter the circulation of the mother and thus, if they are Rh+, immunize her if she is Rh−. The leakage is especially pronounced during the last stage of delivery, when the pressure of the uterine muscle is great.

This last fact explains why the first Rh+ child of a couple in which the mother is Rh− and the father is Rh+ may be perfectly normal. Troubles usually begin with the second Rh+ child and become worse with subsequent ones. It is primarily at the delivery of her first child that the Rh− mother becomes immunized against the red cells of her Rh+ baby, and therefore she may be rich in antibodies when the second Rh+ fetus is developing.

What can be done in practice? If the child is born alive but there are signs that he may have severe damage caused by maternal antibodies, these antibodies must be washed out of his body by special exchange transfusion techniques. Recently another technique has

84 been introduced which prevents the immunization of the mother and thus can eliminate most of the damage due to Rh incompatibility. As we have already mentioned, the majority of first Rh+ children born to Rh− mothers are normal. Immediately after birth, the mother must be treated with anti-Rh+ serum. This will agglutinate all the red cells of the child which may have gone into her circulation and, for reasons that are not entirely understood, will prevent her immunization against Rh. Thus the next Rh+ child will develop in an immunologically virgin field, and will not be affected. It should be added that a woman who is Rh− and has married an Rh+ husband who is heterozygous (*Rr*) can expect half of her children to be Rh−, and will have no problems with such children.

It may be interesting to give some figures on the frequencies of the various types. Approximately 16% of individuals in a Caucasian population are Rh−. The proportion is the same in men and women, of course, since this is an autosomal gene. All remaining individuals of Caucasian origin are Rh+. As computed from Hardy-Weinberg equilibrium, the 84% of individuals that are Rh+ are in part homozygotes (36%) and in part heterozygotes (48%). Thus there is a slightly higher chance that an Rh + person is a heterozygote than a homozygote.

For reasons that we do not understand, there is a great deal of racial variation with respect to the Rh character. Among Africans the frequency of the *r* gene is much lower and only 1 or 2% are Rh− individuals; thus the problem of Rh incompatibility is much smaller among Africans. Among Orientals the frequency of *r* is zero, and therefore this problem is absent.

The type of interaction we have described is often called incompatibility. It is found also with other alleles of the Rh group and with other blood-group systems. In particular, it may be found with A children of O mothers (although it is then clinically different from the Rh type of incompatibility). There are also interesting relationships between incompatibility for ABO and for Rh which we shall not discuss. In any case, the Rh incompatibility is clinically the most important one and, for this reason, in all developed countries all women are tested for Rh type before they give birth to a child.

5.4 HLA and Transplantation

Organ and tissue transplantation has been a major development in surgery during the last decade and, especially for kidneys, a degree of success has been achieved. New and difficult genetic problems are posed, however, by the fact that individuals turn out to be very different from one another at an immunological level. A previously unsuspected variety of immunological differences must exist, and we know that these differences are genetic because transplantation is always successful if made between identical twins, who have perfectly identical genotypes. Transplantation from one individual to another taken at random from the population is usually not successful, or only moderately successful, even if one tries to minimize the immunological reaction by all the means known to modern medicine.

It is clear today that many different genes participate in the diversity which creates the incompatibility of transplanted organs. For example, ABO has been shown to have a part. However, it is believed that a major role is played by one system of antigens which has been observed mostly on lymphocytes, but is certainly also present on many other tissues, though not on red cells. This system is called HLA; the first two letters are an abbreviation for human lymphocyte, and A stands for the first (and thus far the only) clearly described system. At least four closely linked genes are known, and they have been assigned a chromosome number. Each of these genes has several different alleles. The number of different combinations for this system is so great that the chance of finding two unrelated individuals identical for HLA is very small. Closely related individuals are, of course, more likely to be similar and thus can more easily meet the criteria for a sufficiently close immunological match to make successful transplantation likely.

If the recipient contains all the antigenic substances present in the donor, it will not develop antibodies against the grafted tissue or organ. There is another possibility, though perhaps somewhat less important. The transplanted organs also may have antibody-making cells, and these may develop antibodies against the host if the host has antigens which the donor does not have. Ideally, therefore, there should be a complete genetic identity for all antigenic substances which can develop incompatibility upon transplantation. This identity exists only among identical twins, but few people who need a transplant have the advantage of having an identical twin ready to donate an organ. For practical purposes, the rule is that transplants between full sibs (siblings) are likely to be better than transplants between two individuals taken at random or even between parent and offspring. Consider a mating between two individuals $A_1A_2 \times A_3A_4$, where A_1, A_2, A_3, and A_4 indicate four different alleles for the

86   HLA system. These individuals can have children A_1A_3, A_1A_4, A_2A_3, and A_2A_4 in proportions of 1/4. Any transplant from parent to child or vice versa is incompatible for one allele; but any child has a probability of 1/4 of being identical to a sib for HLA.

An important series of recent discoveries has shown that several diseases are in close association with HLA alleles. In a few cases, the association is so close that the HLA system can be used for diagnostic purposes! Thus an ailment of the vertebral column, ankylosing spondylitis, is found in 90% of the patients associated with a particular allele of an HLA locus. Reasons for these associations are not entirely clear, but may depend on very close linkage of the HLA genes with neighboring genes that have some specific immunological functions, like the capacity to respond to certain groups of antigens.

PROBLEMS

1. Can a child of the blood groups given below be born in the given mating?

	Child	Mating
		♀ ♂
(a)	O	A × A
(b)	O	AB × A
(c)	A	B × B
(d)	A	B × AB
(e)	A	O × O

2. If an Rh− woman marries a man of unknown Rh genotype, what is the probability that she can only have Rh+ children with that husband (i.e., that the husband is homozygous for Rh+)? Assume that the frequency of Rh+ in the population is 84%.

3. In the mating of a heterozygous Rh+ husband and an Rh− wife, 1/2 of the children will be Rh− and therefore unaffected by Rh hemolytic disease. With a 16% frequency of Rh− in the population, what is the proportion of Rh+ husbands that is heterozygous and therefore can have Rh− children?

4. In Problem 1, some children are incompatible with the matings. A "suppressor" gene is known for ABO. This is an autosomal recessive, which, when homozygous, makes an individual that would normally express genes *A* and/or *B* seem more like an O individual, and therefore he is bloodtyped as an O. Which of the incompatibilities detected in Problem 1 might be thus explained?

Chapter six

Identical twins are usually so similar that they give an indelible impression of the power of inheritance. But the phenotype of each individual may also be deeply affected by the environment in which the individual develops. Note that some traits, like the attachment of hair in these twins, may be mirror images. (Courtesy M. Feldman.)

Characters Affected by Inheritance and Environment

Many traits are affected so deeply by "environmental" effects, and possibly also by a variety of genes, that it may be very difficult to fit them into a simple Mendelian scheme.

6.1 Continuous and Discontinuous Variation

What exactly do we mean when we say that a character is inherited? Obviously we refer to *biological* inheritance, not *social* or *cultural* or *economic* inheritance, although the latter are very important influences on behavioral traits and may also have some long-term effect on biological traits. The best way to answer the question is to go through the operations necessary to determine whether or not a certain character is inherited. We usually study one character at a time, carefully defined, and much of our subsequent work depends on the nature of the character and the mode of its variation. It is especially important to distinguish two major categories of characters: those that vary continuously and those that vary discontinuously. By *continuous* variation, we mean that the character can take any value between two extremes.

90 An example is height, another is weight, and yet another is IQ. Obviously there are very many such characters. Those that we have mentioned are all measured in standard ways, and the first two are very easy to measure. However, there are other characters that vary continuously but are difficult to measure in a satisfactory way—for instance, the shape of the nose. Obviously the nose can be measured, but if we really want to define a nose we have to take a fairly complex set of data into account—width, length, slope, curvature, etc.—and we may never be fully satisfied that the data provide a complete and accurate description. Characters of such a complex nature are clearly the most difficult to study.

When we speak of *discontinuous* variation we are referring to characters that exist in two or more alternative forms which are clearly distinguishable from one another, without any overlap between the various types. Examples we have seen are blood groups, electrophoretic differences between enzymes, and proteins in general. It should be intuitively clear that variation of a discontinuous nature is much easier to study than continuous variation. This was clear to Mendel, who explicitly stated in his paper that he chose for his analysis characters showing discontinuous variation; this choice was one of the reasons for his success. Even with discontinuous variation, however, difficulties may be encountered and the strength of evidence for the inheritance of the character may depend on the depth at which the investigation was carried out and the nature of the data collected.

There are over a thousand characters, known today in man, inherited strictly according to Mendelian rules. Many of them are rare diseases, and many are traits that are, so far as we know, of no importance to our health or general behavior; a catalog of them has been published by McKusick. We know something about the physiology of some of these characters, and for a small number of them we can even specify which protein is affected and how. In some cases we can trace almost the complete history of a condition from the DNA to the phenotype and the evolutionary history. The classic case is sickle-cell anemia. Unfortunately we still do not know an effective cure for sickle-cell anemia, but the knowledge we have accumulated will permit progress in the development of one. Superficially, it may seem surprising that for some other genetic diseases such as hemophilia and some disorders of endocrine glands a therapy is already available, while for sickle-cell anemia, which is so much better known genetically, no satisfactory cure is available. It should be noted, however, that hemophilia and other inborn errors of metabolism belong to the class of diseases in which a protein is missing, and therefore therapy can take the form of administering the missing protein directly to the patient. The cause of sickle-cell anemia is not that a protein is missing, but that a different one is present, and no simple substitutional therapy is thinkable. Moreover, this protein is not in the fluid part of blood, but inside red cells. However, with the rapid progress of molecular genetics, such a method of therapy should become available soon; some data have been published that give grounds for optimism.

Let us now consider characters that vary continuously but are more complicated than height and weight and IQ, perhaps not so complicated as the shape of the nose, but nevertheless complex. For instance, let us consider the common malformation known as a cleft palate. This is not an example of strictly continuous variation, because a cleft palate is either there or not there. But this malformation certainly exists in various degrees, and in any case analysis has shown that an uncomplicated, single-gene interpretation is not valid. What can we do with such a character? The explanations that we have before us are of two kinds. It may be that more than one gene—perhaps many—affects the character in question, and therefore the rules of inheritance are complex; we will call this the polygenic hypothesis. Another explanation is that the character is affected by environmental variables as well as by genes; let us call this the environmental hypothesis. For instance, infections may have an effect, and these will vary in the histories of different individuals. Or perhaps nutrition is important —and it is hard to deny that this is true in the case of weight (to some extent, undoubtedly, height is also affected by nutrition). Many other factors can influence the characters we study, especially if these are behavioral characters such as IQ. The ensemble of all these factors that influence the development of an individual, from the stage of a single zygotic cell onward, is called the *environment*.

The two hypotheses, the polygenic one and the one assuming that environmental factors also play a part, do not exclude each other. In general, for some characters gene effects may be more important, and for others environmental influences may be more important.

6.2 Twins

The simplest way to see whether a character has some heritable component is to study the *similarity between relatives*—the closer the rela-

92   tionship, the greater must be the similarity. Sir Francis Galton started this approach during the last century; he also introduced measurements of similarity which are similar to those commonly employed today and which in statistics are called correlation coefficients. Galton also realized that twins can be of great value in the attempt to determine whether a character is inherited or not.

There are two types of twins, namely identical twins and nonidentical or fraternal twins. They are often referred to as monozygotic and dizygotic twins, or MZ and DZ twins. These two terms describe the genetic explanation for the difference between the two types of twins. Identical twins come from a single zygote, that is, from a single egg fertilized by a single sperm. The zygote then divides into two equal cells, each of which develops into an embryo. Thus monozygotic twins are absolutely identical genetically and represent a sort of (almost) ideal experiment of nature to test whether two different individuals who are genetically identical may develop differently. Clearly, if genes are paramount in determining a trait, that trait will be very similar in two MZ twins. Dizygotic twins come from two independent zygotes, that is, two eggs, each of which has been fertilized by a separate sperm. They are therefore expected to be as similar or as different as any two sibs, apart from the fact that they are of the same age and have developed together at the same time both in the uterus of the mother and afterwards.

Twins are not infrequent. At least among Caucasians, twins represent a little more than one in every hundred births. Triplets are much rarer, and quadruplets, quintuplets, etc., are rarer still. Among Caucasians, about one twin pair out of three is of the monozygous type and the rest are dizygous. In other racial groups the ratios are somewhat different. Among Orientals they are approximately reversed. Among Africans the frequency of MZ twin births is about the same as in Caucasians, but that of DZ twins is higher. We do not know exactly what determines twinning, especially monozygous twinning. Here we are interested only in the use of twins for the purpose of establishing whether or not a character is inherited.

The critical reader will want to know how we can distinguish monozygous twins from dizygous twins. Monozygous twins, being genetically identical, must be perfectly identical to one another for all those traits that we know are genetically determined. A very sensitive test would be to transplant the skin of one onto the other. In MZ twins the graft will be accepted; in DZ twins the graft will

practically always be rejected. But this test is, for practical reasons,
almost never resorted to. All other genetic characters that can be tested more easily are used—sex, blood groups, protein differences known to be genetically controlled, etc. If twins of a pair differ with respect to one of these traits, they can be considered DZ. There is a small probability of including among those twins diagnosed as MZ a few DZ twins that just happen to be identical for all characters tested. The probability of such an error can, however, be computed and reduced to a very small value, if enough characters are used for the diagnosis.

6.3 Testing for Inheritance of Quantitative Characters

If we measured the height of a number of pairs of identical twins, we would find the mean difference between members of a pair to be about 1.5 cm. Thus MZ twins are not identical for height, but they are close. In dizygous twins the mean difference is about 4.0 cm, which is higher than in identical twins. The mean difference of height between individuals of the same sex taken *at random* from the population is about 6 cm. With these figures as evidence, we can say that the effect of inheritance on height must be great, because MZ twins are so much more similar than are DZ twins or random pairs. There also must be some environmental effect, because MZ twins are not identical for height. There is also independent evidence for environmental effects on stature. For example, if we look at the mean stature of people in developed countries during the last 50 years, we see that there has been a tremendous increase—5 or more cm in 50 years. Improved nutrition and other changes in health and living conditions have almost certainly contributed a great deal to this change. This environmental effect (due to change of nutrition, etc., with time in populations of European origin) is *not* necessarily represented in the differences between two identical twins. Moreover, the same type of environmental change that caused an increase of stature in the U.S. and Europe may not take place at all in other populations. The "environment" is very complex, and even in the case of the stature increase, we do not know exactly what caused the change.

Let us consider a character of greater social significance than height, such as IQ. Intelligence quotient is a measure of intellectual performance in some standard tests, which involve reasoning ability,

94   memory, and knowledge of the language and of the culture to which an individual belongs. (The ideal of psychologists is an intelligence test that is "culture-free," i.e., does not depend on the culture in which an individual has developed, but as yet this ideal has not been achieved.) IQ is measured on an arbitrary scale such that the average between individuals is 100 IQ points and variation is such that 95% of individuals are included in the range from 70 to 130. The other 5% are almost equally divided between those above 130 and those below 70. Averaging observations made in Great Britain and the U.S., we find that MZ twins have an average difference of about 6 IQ points and DZ twins have an average difference of 11. Random pairs show a difference of 21 points. More or less the same conclusions can be drawn here as in the case of height. Clearly, inheritance can be important, because MZ twins are more similar than DZ twins, and these are more similar than random pairs, but MZ twins also show differences between one another. Therefore, some environmental effect must also be present, and it appears, from the two sets of figures, that the environmental effect on IQ is a little greater than that on height.

There are, however, some logical traps in this kind of comparison. Identical twins, because they are identical and because of other reasons, are usually exposed to (and even tend to choose) a more similar environment than are nonidentical twins. Parents may contribute to this effect. For instance, identical twins are frequently dressed identically by parents while the twins are young, more so than DZ twins. The famous human geneticist L.S. Penrose said, only half-jokingly, that if one looked at twin data uncritically, one might conclude that clothes are inherited biologically. One possible way out of this trap is to make use of the fact that in some cases twins are separated soon after birth and are raised in different families. For the two characters that we have just mentioned, height and IQ, it can be seen that the mean difference between MZ twins when they are brought up apart is somewhat larger than that between MZ twins when they are brought up together. Therefore, the differences between the environments offered by different families may have some effect on these traits.

Actually, there are other, more direct proofs, that environmental effects on IQ cannot be negligible. Adopted children raised in "good" families seem to show a considerably higher IQ, on average, than what might otherwise be expected. Sibs show substantial effects of birth order and family size: on average, later-born sibs have lower IQ's than those born earlier in the same family. The magnitude of the birth order effect should not be exaggerated; it is not large compared with the individual variation between sibs. However, these and other well-ascertained facts show that differences in the environment can easily be responsible for substantial IQ differences. The overall data are compatible with both hypotheses, namely that both genetics and environment play an important role. We will consider briefly, at the end of this chapter, attempts at measuring their relative roles.

6.4 Behavioral Disease: Schizophrenia

Let us now study in more detail an example of considerable social importance, schizophrenia. This mental disease is very widespread; indeed, almost 2% of all individuals born in most countries and racial groups are affected by it. There are many forms of schizophrenia, but usually a schizophrenic is either very excited or devoid of interest in the outside world. Occasionally he may be dangerous to society and to himself. He does not lose his intelligence, but simply does not use it because of a lack of interest. He may have hallucinations and other symptoms.

To indicate the importance of schizophrenia, one can note that about half of all the beds in psychiatric institutions in the United States are occupied by schizophrenics. We know practically nothing about the biochemistry of the disease. Much discussion has centered on the question of whether schizophrenia is inherited or not. There are certainly various degrees of the disease, and there are certainly many individuals in the general population who do not develop the disease but have slight schizophrenic tendencies, which may not seem especially dangerous or require treatment.

If we look at cases of schizophrenia among twins, we find that when one of two identical twins is schizophrenic, the other is also schizophrenic in 38% of the cases (this figure is an average over many investigations). If schizophrenia were entirely inherited, this percentage would be 100%. Thus it is clear that environmental differences must play a substantial role in the development of individuals. Is there room for a genetic component when we find such a low concordance, as it is called, between identical twins for schizophrenia? There is room, because we find the corresponding percentage in DZ twins to be only 12%, which is decidedly less. This value is fairly close

96   to the rate of concordance found among ordinary sibs, as would be expected, since DZ twins have the same degree of genetic similarity as do ordinary sibs. Practically all investigations have shown a higher rate of concordance for schizophrenia among MZ twins than among DZ twins. It seems likely that inheritance plays some role in determining who is afflicted by the disease. But the evidence from ordinary identical twins (who are usually reared together) is not very strong, for we know that twins share a common environment and also may influence each other to a large extent.

A few cases are known in which MZ twins that were concordant for schizophrenia were brought up apart. These cases seem to reinforce the conclusion that inheritance is important in the determination of schizophrenia, but they are too few to rely on. However, another possible approach exists, the study of children born to schizophrenic parents but *adopted* into normal families. Studies of this kind have shown that such children tend to develop schizophrenia more than do controls, that is, adopted children born to non-schizophrenic parents. Adoption studies have thus confirmed that genetics does indeed play a role in schizophrenia, but the low concordance found in MZ twin studies also shows that environment plays a role. Thus an individual may be genetically more likely than another to develop schizophrenia, but he may or may not develop it, depending on the environment in which he was raised. Unfortunately, we still do not know which of the many elements in the environment play a role in the development of this disease.

6.5 Heritability

Many attempts have been made to express in quantitative terms the relative roles of environment and genetics in the determination of characters of the types just described. Typically these attempts involve the computation of a quantity called "heritability," which should express the relative role of genetics as a percentage of the total effect of genetics and environment. It is unfortunate that a full understanding of this quantity requires relatively advanced knowledge of statistics, so that a complete analysis of the concept and its limitations would be out of place in an elementary introduction. An article in *Scientific American* and two books by the author, in collaboration with Walter Bodmer (cited in the references), may be consulted for more complete treatments.

Suppose that the IQ of an individual could be construed as the sum of two quantities: $IQ = G + E$ where G is the contribution from genetics and E is that from environment. The question that heritability tries to measure is the relative importance in the *individual variation* of G and E, but it does *not* answer the problem: what is the relative magnitude of G and E? A common source of confusion is that uninformed people may tend to interpret heritability as meaning just that—namely, relative magnitude—but in fact one is limited to asking: how much does the G component *vary* from one individual to the other, and how much does the E component vary?

For IQ, heritability is the ratio of the variation of the G component of IQ to the total variation of IQ (always between individuals of the population being studied). This may seem highly indirect, and it is, but it is a question that under some conditions might be answered reasonably well. Even in this simple form, and assuming that simple conditions hold, however, there are formidable problems of estimation from the available data. Moreover, most methods of estimation tend to assay a different quantity. Estimates that have been offered for IQ vary from 45% to 85%.

In any case, it is unavoidable to conclude that estimates of heritability of IQ in human beings have very little to offer for making useful predictions. IQ is just a narrow facet of human personality, but still a conglomerate of different behavioral activities, strongly influenced by cultural values and teachings. The validity of heritability estimates is limited to the population studied, the time it is studied, and the methods employed. Such estimates are far from being universal quantities. They only measure what variation exists at the time in the E and G components, but they contain basically no information on how given genotypes may behave in other environments which are not tested or which are uncommon in the population examined. Not only do estimates have no use for individual prediction, but they do not help predicting how the average IQ of a population would change under some environmental change. They can be widely misinterpreted, as happened in the recent discussion on IQ and race. It is known, and has been widely publicized, that there is a difference between the average IQ of U.S. whites and blacks. There are also IQ differences, on average, between social strata of Western societies that have been investigated. The inference was made that, because heritability of IQ is high, these differences are likely to be genetic. But this inference is invalid. Even if it were correct, the widespread prejudice

98   that genetic gaps cannot be filled is wrong. Lack of appreciation of these basic scientific points, and the social and political implications behind these statements, have helped turn this issue—one which scientifically is neither too exciting, nor likely to be solved to everybody's satisfaction by our present means of analysis—into a nationwide turmoil. We will not examine the social and political implications, some of which may be easily perceived by everybody. As to the prejudice that a genetic gap is "incurable," it may just be worth remembering that the phenotypic effects of PKU, a clear-cut genetic defect which reduces IQ from 100 to below 50—a monstrous fall—can be totally eliminated by administration of an appropriate diet in the early years of life. This was made possible by the understanding that in PKU children phenylalanine levels are too high because of a genetic block in the conversion of phenylalanine into tyrosine. High phenylalanine levels are toxic for the developing brain, but they can be lowered, in the presence of a genetic block, by reduced administration of phenylalanine. This at once points to ways of beneficial scientific developments: a closer analysis of the interaction of genotype and environment, which can only profit us more by being explored also at the physiological and biochemical level.

PROBLEMS

1. A sib of a person born with a harelip or cleft palate (a relatively rare defect) has a 3.5% chance of having the same defect. What would be the chance if a regular recessive gene were responsible for the defect?

2. Suppose that when one member of an MZ twin pair stutters, the other twin has the same defect 40% of the time. (a) What frequency would be expected if the trait were determined entirely genetically? (b) What conclusion can you draw from this single observation? (c) What additional conclusion would you draw if you knew that DZ cotwins of affected individuals have a definitely lower incidence than MZ cotwins? (d) What observations, if at all possible, would you like to have in order to strengthen your conclusions?

3. Smoking seems to "run in families," in the sense that relatives of smokers smoke more frequently than those of non-smokers. What kind of observations can show if predisposition to smoking is truly genetic and not due to some kind of imitation, suggestion, or other kind of psychological contagion from relatives?

Chapter seven

Races were the outcome of separation and subsequent geographic isolation which lasted for some tens of thousands of years. During this time there was opportunity for differentiation. Differences that we observe with the naked eye are mostly due to skin color and facial traits. They most probably represent adaptation to different environments (of which climate was certainly a very important part). A random sample of genes shows far fewer differences on average. The young women in this photograph are Aka Pygmies from Central Africa.

How We Change, How We Differ

All living organisms evolve, that is, they change generation after generation, very slowly. The differences between "races" are part of this process.

7.1 Mutation

We have already mentioned that errors in the process of copying may occur, especially when cells, and therefore their DNA, reproduce. Such an error is usually the insertion of one nucleotide in the place normally taken by another. The result may be the substitution of one amino acid in the protein made by the segment of DNA concerned, the exact consequences being predictable by the genetic code (see Fig. 7.1,*a*). Another possibility is the loss or addition of a base (adenine, thymine, cytosine, or guanine) on the DNA molecule. Since the genetic code operates in terms of triplets of bases (see Section 2.4), the loss or addition of *one* or *two* bases will give the sequence of bases an entirely different meaning from that point on (Fig. 7.1,*b*). The

101

a. A substitution of one nucleotide in DNA (underlined) may determine the substitution of one amino acid in the protein.

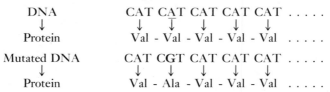

b. Insertion of a nucleotide changes the whole sequence of amino acids from that point onwards.

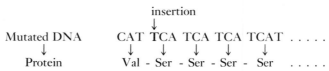

c. Insertion of three nucleotides corresponds to the addition of an amino acid.

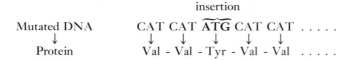

Figure 7.1
Mutation as a change in DNA and its effects on protein. We assume for simplicity that the DNA fragment considered is highly repetitious, having the same triplet repeated in sequence (see the genetic code, Section 2.4).

sequence of amino acids, therefore, will be different from the correct sequence, and the protein will likely be nonfunctional. This type of mutation, called a "frameshift" mutation, has also played some role in evolution. A third possibility is the loss or addition of *three* consecutive bases. In this case one amino acid will be deleted from or added to the protein chain (Fig. 7.1,*c*).

Other types of mutation may also occur. Chromosomes may break and rejoin in different positions, they may be lost, or sometimes duplicated. Mutations of these types, called chromosome aberrations, usually have more serious consequences than do the smaller mutational changes mentioned above (gene mutations).

7.2 How Frequent Are Mutations?

We intuitively expect that mutations are not very frequent. The human organism is like a complex piece of machinery (much more complicated than a missile), all of whose parts, or at least all vital ones, must function. In some cases the substitution of even a single amino acid in one protein may be lethal to the organism. Not all mutations, however, have such drastic effects. Some cause little or no disadvantage to their carriers, and some may even be advantageous, as we shall see when we speak of natural selection.

Mutation rates in man, as in other animals and plants, can be estimated. For example, the dominant mutation leading to the type of dwarfism known as achondroplasia appears once in every 20,000 gametes. This is one of the highest mutation rates known. Another frequent mutation is that leading to hemophilia, the coagulation disease, which, as we have seen, is sex-linked. The rate of that mutation is about one in every 50,000 gametes. These two examples have been chosen from among the highest rates; the average mutation rate *per gene* in man is likely to be lower than one in every 1,000,000 gametes. Then in every generation one gene of a given individual has a chance of about 1/1,000,000 of being mutated. But since there are many different genes in man, perhaps as many as a million, every gamete may contain a new mutation of one of its genes. These estimates all refer to what are currently called spontaneous mutations, that is, mutations that occur in the absence of any specific (additional) outside agent.

What causes mutations? X-rays and all types of ionizing radiation from cosmic rays and all sorts of particles emitted in radioactive decay are known to produce mutations. The introduction of atomic energy has increased the amount of radiation we are all subjected to, and will further increase it. The use of X-rays for medical purposes has also contributed heavily to the increase. What fraction of the entire mutation rate can be attributed to this radioactive background? We still cannot give exact figures for man, but we estimate that, of the total "spontaneous" mutation rate, the fraction that is due to the radioactive background is somewhere between 3% and 30%. One important consideration here is that there is no threshold below which radioactivity is completely inactive in producing mutation. Therefore, any man-made addition to the radioactive background will increase the mutation rate. We know that many of our medical problems are due to deleterious mutations, and therefore we would like to reduce rather than increase the mutation rate. It was for this reason that human geneticists were active and effective shortly after World War II in pleading before the United States Congress for a reduction in or abolition of atomic bomb testing.

However, it seems almost certain that radioactivity does not explain all of the mutation rate observed. What is the rest due to? Probably chemicals that enter into the composition of our food or atmosphere have a part. Clearly, study of the possible mutagenic action of substances that are widespread throughout the world, such as pesticides, is of considerable social importance. Even some viruses have been charged with mutagenic action, especially chromosome changes.

A search for chemical mutagens is a useful part of any program to improve the environment. Mutagens may have detrimental effects not only on our descendants by the mutations they produce in our gametes, but they can also be deleterious to us personally, for they may induce mutations in our non-gametic (somatic) cells. Some of these mutations may be responsible for cancers. Thus, it is not too surprising that many substances with mutagenic action are also endowed with carcinogenic (cancer-determining) action and vice versa.

7.3 Natural Selection

During the second half of the last century, Charles Darwin was the greatest exponent of the theory that natural selection determines the course of evolution and the adaptation of organisms to their environment. His battle was a long one, and even today there are a few people who do not believe that evolution takes place—mostly for reasons of religious fanaticism or misunderstanding. Natural selection can be expounded today in mathematical terms, and could be (in fact, has been) given the form and strength of mathematical theorems. For reasons of simplicity, we will avoid here any quantitative treatment, which can be found elsewhere.

Let us assume that one gene has mutated, and we know that in this case the change is potentially transmissible to all the descendants of the organism that carries the mutated gene, because the mutated gene is now the master copy from which all future copies will be made, and these future copies will appear in the descendants. The fate of a mutated gene, which we shall call the *new* allele of a given gene, depends on the effect this allele has on the organism that carries it. Let us assume that the change happens to be advantageous to the individual, at least in the particular environment in which it lives. We can cite one such example in man, that of G6PD deficiency in a malarial environment. In one case this gene has risen from a frequency that must originally have been very low, to at least 60% (among Kurdistani Jews); among other groups, the frequencies reached are smaller. It is possible that, if malarial conditions had continued, the Kurdistani Jews may have become 100% G6PD deficient. From historical data we can approximately establish the time over which the observed change in gene frequencies occurred; the period must have been of the order of 2000 or 2500 years. This is perhaps the most rapid genetic change in man about which we have some detailed information. Similar examples of almost equally rapid evolutionary changes all involve genes that, like G6PD, determine a greater resistance to malaria (like sickle-cell anemia or thalassemia). It should not be surprising to find that the known instances of rapid evolution by natural selection all occurred in the context of malaria. This disease has been one of the worst killers in all tropical regions, as well as in some temperate ones, those in which the mosquito carrier of the disease (*Anopheles*) can survive and prosper.

As some simple mathematical computations would show, we cannot expect such rapid evolution for most other genes. In fact, a widespread tendency today is to assume that many—but certainly not all—single amino-acid substitutions observed in proteins are not very meaningful from the point of view of natural selection. Either they are completely neutral, bringing absolutely no harm or benefit to the carrier, or the selective advantage or disadvantage is so small that it cannot easily be measured. This "selective neutrality" cannot apply to all genes, of course; otherwise it would be impossible for organisms to adapt to their environment. Adaptation involves an increase in the number of individuals who are more fit for their environment over those who are less fit, and this requires at least some of the mutations to be advantageous.

The fraction of mutations that benefit the carrier may be relatively small, but without it there would be no adaptation to the environment. On the other hand, there is considerable evidence that a large fraction of all mutations is *deleterious* to the carrier. Here we have to remember the fact that we are *diploid* organisms, that is, all our autosomes are in duplicate. When a completely new mutation arises, it will usually arise in one chromosome of a gamete, which will fertilize another gamete. This other gamete will probably carry the normal rather than the mutant gene, and thus the zygote formed will be heterozygous for the mutant allele. Thus, if the new mutation is recessive, it will be unobserved because the individual heterozygote for the mutation will

106 be phenotypically normal. The fact that we have a double set of chromosomes does help to hide the deleterious effects of those mutations that are recessive. This gives us a clue to the strategy that nature may have adopted in making many organisms diploid. It is the same strategy that airplane designers have used. A single-engine plane whose engine failed in flight would, with high probability, crash. To reduce the chance of a crash, a single-engine plane usually carries two carburetors, two magnetos, and a number of other duplicate devices. Even greater safety is offered by having two or more engines instead of one. Diploidy serves the purpose, among others, of screening us against the dangers of recessive mutations. It cannot help us, however, against dominant deleterious mutations, but these, as we shall see, are quickly disposed of by natural selection.

7.4 Rarity and Multiplicity of Genetic Disease

A *dominant* deleterious mutation appearing in one gamete shows phenotypic effects in the zygote that arises from that gamete. If the individual carrying the mutation dies or does not reproduce, the newly mutated gene will disappear. For example, the chance that an achondroplastic dwarf will leave progeny is only 20% of that of a normal individual. The quantity, 20%, is called the *fitness* (of the heterozygote for achondroplasia); its complement, 80%, is called the *selective disadvantage*. However, some of these dwarfs do reproduce, and they will transmit their gene to their offspring. In addition, with every new generation, new mutations will appear at the relatively high rate we have already mentioned. Thus a certain number of dwarfs will be present in every generation. Two forces are acting in opposite directions: mutation producing the new allele, and selection weeding it out. An equilibrium will be reached when the two forces are perfectly equal.

We can visualize the action of these two forces by imagining a bag of beans, most of which are black, but to which we keep adding, at every generation, some white beans. The white beans represent mutations. On the other hand, at every generation we also remove some white beans; this represents selection against the mutant. The number of white beans in the bag will not change when the number that we add (mutation) is equal to the number that we subtract (selection). When selection against the mutant is very strong, as in the case of a highly deleterious gene, the number that we subtract at every generation almost equals the total number present, and most of these will have arisen *de novo*, that is, by fresh mutation. Thus the frequency of individuals affected by a given dominant genetic disease that is highly deleterious (has a low fitness) will be of the order of the mutation rate. This mutation rate varies a great deal from gene to gene, but usually any given genetic disease will be rare, because mutation for a given gene is usually rare.

So far we have discussed dominant deleterious mutations. What about recessives? Because they can hide in heterozygotes, it is possible that new mutations will keep accumulating until the recessive deleterious mutants reach relatively high frequencies. It is only when two heterozygotes for the same mutant allele mate that the homozygous condition can be found among the progeny, and even then only in 1/4 of the progeny. Only at this time does a recessive deleterious gene become exposed to natural selection, and because it is deleterious, it is weeded out. Thus it is to be expected—and simple computations confirm this expectation—that the frequency of a recessive deleterious gene can become fairly high in a population. We have seen this in the case of phenylketonuria. However, the frequency of affected individuals will still remain very low in the population. As can be seen from the Hardy-Weinberg rule, the frequency of affected individuals is the square of the gene frequency. If the latter is 1% (0.01, or 1 in 100), the former will be 0.01% (0.0001, or 1 in 10,000). If the gene frequency is 0.1%, the frequency of affected individuals will be 1 in a million, and so on. Thus for recessive deleterious diseases as for dominant ones, the frequency of affected individuals will be of the order of magnitude of the mutation rate, unless other complicating factors occur, such as those discussed in Chapter 8.

Generalizing, we can say that, apart from some exceptional situations still to be discussed, every genetic disease is rare. However, there are hundreds of thousands or millions of genes in the human organism, and every one of them has the potential to produce mutants that may be deleterious to the organism. Therefore, there may be millions of different genetic diseases, and the total toll of genetic disease may be very high when we sum over all the many possible diseases, even though each is rare. Today we know of only about a thousand such diseases (and our knowledge of most of them is incomplete), but this number is bound to increase.

108   7.5 Duplication of Genes

Among mutations we have not yet discussed there is one which is of special importance in evolution. This is the duplication of whole genes or segments of chromosomes so that there are several copies of that gene in one chromosome (or in the members of a different chromosome pair). We have direct evidence that this must have happened repeatedly, and that there are many duplicated genes, usually close to one another but often also on different chromosomes, which make very similar proteins. The similarity is very close, sufficient to convince us that both genes had a common origin.

The classic example is that of the hemoglobins. We know at least five chains of hemoglobin in mammals; they are named alpha, beta, gamma, delta, and epsilon. We know that the ordinary hemoglobin of an adult is made of two chains of alpha and two of beta. But in early fetal life, alpha and gamma are used, and this second type is then gradually replaced by the normal adult alpha-beta type. We know that the alpha chain is made by a gene on one chromosome, and that beta, gamma, and delta are made by three genes, all closely linked on another chromosome. We can also determine approximately when the duplications took place by studying the presence of these molecules in the various vertebrates. For instance, the delta chain arose only relatively recently (see also Fig. 7.2). The delta chain differs from its prototype, the beta, by only five or six amino acid substitutions. In general, the older the separation between two duplicated genes, the greater the number of amino acid differences between them. Figure 7.2 shows the numbers of amino acid differences for the main hemoglobin chains. It may be added that there is another protein, myoglobin, which serves a function similar to that of hemoglobin. However, it is located in another tissue, that of muscle. Myoglobin is sufficiently similar to hemoglobin that the two molecules must have had a common origin. Since myoglobin shows even more amino acid differences than do the hemoglobin chains shown in Fig. 7.2, the separation of myoglobin from hemoglobin must have occurred earlier than the various duplications of the original hemoglobin chains.

There are many other cases in which it can safely be assumed that a gene has duplicated several times and each duplicate has evolved independently. It is interesting to speculate how the duplications may arise. Probably the simplest mechanism is that of illegitimate crossing-over, illustrated in Fig. 7.3. In general, crossing-over is a

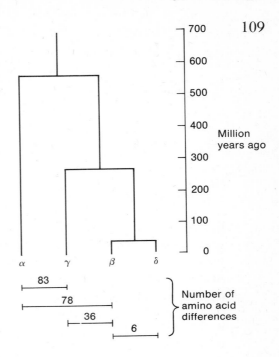

Figure 7.2
Molecular evolution of hemoglobin chains α, β, γ, and δ, and estimated times of evolutionary separation. (From a table by E. Zuckerkandl and L. Pauling in V. Bryson and H. Vogel, eds.: Evolving Genes and Proteins, New York, 1965, Academic Press.)

very precisely controlled process which is believed to be a consequence of the tendency of strictly similar chromosomes to pair with each other in a very exact way. Illegitimate crossing-over may occur if DNA segment *de* is sufficiently (perhaps because of an old duplication) similar to segment *fg* (Fig. 7.3) that *d* can pair with *f* and *e* with *g* (normally *d* would pair with *d*, *e* with *e*, *f* with *f*, and *g* with *g*). The consequence of such crossing-over is that one of the two strands formed contains a duplication, and the other a deletion (that is, a missing segment).

Gene duplication is a very powerful factor in the creation of more and more complex organisms. The fact that the same protein is made by two different genes gives freedom to one of the two genes to evolve in various ways and perhaps to take on new functions. This was certainly the case, for instance, with myoglobin and hemoglobin, which must have originated from a duplication early in the history of chordates. In the beginning, the duplicate gene may remain without a specific or important function. For instance, there is no very important function attached to the delta chain of hemoglobin (so far as we know today), which has a rather recent origin. It is used only to make about

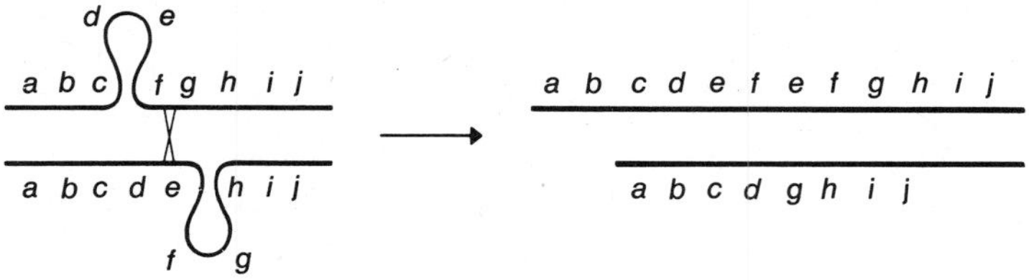

Figure 7.3
Illegitimate crossing-over, leading to duplication and deletion.

2% of the hemoglobin in the adult (called hemoglobin A_2, which consists of two alpha chains and two delta chains). We could probably lack hemoglobin A_2 entirely and still be perfectly normal. But a time may come, as a result of change in environmental conditions, when this protein reveals unsuspected advantages. It is in this way that living organisms develop a great variety of potential mechanisms, which are chosen by the process of natural selection whenever environmental conditions present the opportunity or necessity to do so. Thus chance, in the form of mutation, and choice, in the form of selection, have led to the great variety of complex organisms that populate the earth.

7.6 Other Factors of Evolution: Random Genetic Drift

We have mentioned chance, in the form of mutation, as a factor in evolution. In fact, the particular changes that mutation may produce in a complex molecular structure are almost random; to that extent, the effects of mutation may seem unpredictable. Moreover, the occurrence of any mutation is a random event with a given probability; here again chance plays a role. There is yet another, entirely different way in which chance is directly involved in evolution.

This third way in which chance enters the evolutionary process is related to the fact that no population, however large, is infinite in size. Given the method of human reproduction, the genes that are passed from parents to children represent a *random sample* of the parental genes. We shall use a concrete example to show how sampling can introduce statistical fluctuations that will result in evolutionary changes of a random nature. We will simulate a population by comparing it, once again, to a bag of beans, some of which are black and some white, representing all the gametes which may contribute to future generations. Let us assume that these gametes are 60% black and 40% white. This means that for the one particular gene that we are considering, 60% of individuals are of type *A* and 40% of type *a*. For simplicity we shall ignore the fact that some of the gametes are male and some female, and that a male gamete must fertilize a female gamete. We shall simply imagine that the formation of adults corresponds to taking a handful of these gametes and putting them together two by two to produce zygotes that will eventually, at the end of their development, become sexually mature and thus form the next generation. When we take a handful of gametes at random, we expect that there will be 60% black and 40% white beans, but we also know that we must expect statistical fluctuations. Thus if we pick out 100 beans to represent the genes of the mature adults that will reproduce to form the next generation, we may find, instead of 60 black and 40 white beans, perhaps 58 black and 42 white beans. There is a theorem (the binomial distribution, discussed in books on elementary probability) that could be applied to express exactly the probability of any possible composition of a handful of beans forming the adult population.

Starting with the adult population (our handful of beans), and ignoring the fact that some individuals may die because of natural selection, we shall again form a pool of gametes. This we shall do by putting our handful of 58 black and 42 white beans into an empty bag and letting them, so to speak, produce gametes again; these gametes will represent faithfully the 58-to-42 ratio. The passage of another generation will be a repetition of the same process. We take a handful of beans to represent the adults. But this time we are taking our beans from a population that has 58% black and 42% white, not 60% and 40% as in the previous generation. The process goes on for generation after generation, and statistical fluctuations due to random sampling of gametes will accumulate. It makes sense intuitively, and it can be rigorously proved, that the larger the sample of beans taken to form the adult population, the smaller the fluctuations will be. We have already seen this principle operating when we discussed sampling error. The fluctuations we are discussing are, again, a form of sampling error. An idea of the random fluctuations that can thus take place is given in Fig. 7.4, which shows actual experiments, two with a small number of beans (a small population, $N = 25$) and one with a medium-size population ($N = 500$). Clearly the smaller populations show larger fluctuations. In one of them, after a few generations, one allele became fixed (that is, reached a frequency of 100%) and the other was lost.

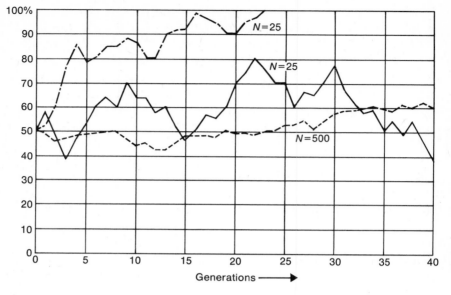

Figure 7.4

The evolution of gene frequencies under random genetic drift, that is, sampling fluctuations due to the finite size of the population. The N-value indicated in each curve is the number of (adult) individuals forming the population. In a larger population the fluctuations are less pronounced, but they take place nevertheless. Independent populations fluctuate independently, and in one case fixation of one allele has occurred. All three populations started at 50% gene frequencies of allele A.

Unless it is reintroduced by some such process as mutation or immigration from the outside, an allele may thus be permanently lost purely as a result of statistical fluctuations.

This process of cumulative sampling accidents has been called *random genetic drift*, or simply drift. It may seem important only for small populations. However, theoretical considerations show that we can expect approximately the same amount of drift in a population of 100 individuals that has drifted continuously for 10 generations, in a population of 1000 individuals that has drifted for 100 generations, in a population of 1,000,000 that has drifted for 100,000 generations, and so on. In other words, population size and time have exactly the opposite effect; therefore, we can expect drift to produce significant differences even in a large population, provided that it has been operating over a sufficiently long time.

In recent years much information has accumulated on the sequence
of amino acids in many proteins, and the same protein has been studied
in a variéty of organisms. If we compare the hemoglobin alpha chain
in various vertebrates, we find that the number of differences in amino
acids is roughly proportional to the time since the last common ancestor. This time is obtained from entirely independent evidence collected by palaeontologists. Such findings have led some authors to
believe that the number of amino acid differences between, say, the
hemoglobin alpha chain of man and that of horse can be utilized as
a sort of evolutionary clock to measure the time that has elapsed since
two organisms diverged in the course of evolution. To be sure, this
clock does not work with the precision that a Swiss watchmaker might
desire. For instance, if we use different proteins as clocks, we find
that they tick at somewhat different rates. A topic of discussion at
the present time is whether the changes observed in the study of evolution at the molecular level are largely or wholly determined by
genetic drift, or whether they represent to a major extent the result
of adaptation due to natural selection. The fact that different proteins
and different amino acid sites of one protein show different rates of
change indicates that the role of selection cannot be entirely dismissed.
The possibility remains, however, that at least a fraction of the amino
acid substitutions observed in molecular evolution can be considered
as practically neutral from a selective point of view, and therefore the
result of genetic drift alone.

7.7 How Many Differences Are There between Individuals?

No person objects to the notion that all humans belong to one single
species. This means that all humans are potentially interfertile and
that there is no limitation to mating between humans or on having
fertile progeny. This does not mean, however, that all humans are
genetically the same. In fact, work in recent years has shown that the
diversity between individuals is much greater than we had previously
anticipated. For almost every gene that has been investigated, there
has been evidence that more than one allele exists, and frequently
more than two. Since there are millions of different genes in one human
organism, the potential number of different individuals is incredibly
high. Only an infinitesimally small fraction of these possible genetic
constitutions can be realized among the approximately four billion

114   people living on the earth today. If any two genetically identical individuals exist, they are almost certainly identical twins. All other individuals differ by a considerable number of genes. This is true even
of persons who are closely related, although an increase in the close-
ness of relationship also increases the number of shared genes.

Such considerations ridicule completely the concept of a "pure"
race. With so many differences between even a parent and a child,
how can one divide the human population into clear-cut groups of
individuals all of whom are very similar to one another? We shall come
to this point again when speaking of inbreeding. Much of the misunderstanding arises from the very frequent confusion by the layman,
sometimes also by scientists, between cultural and biological attributes
of ethnic groups. There may be cultures that are relatively "pure" in
the sense of homogeneous, but today anyone who believes in the existence of biologically pure races in man must be considered a charlatan.

7.8 The Concept of Race and Racial Differences

It is fairly easy to group humans broadly, either by simple inspection
or by detailed genetic tests, into three major classifications: Africans
or blacks, Caucasians or whites, and a third group which is more het-
erogeneous than are the first two and includes all the populations living
in the Pacific area (Orientals, including Chinese and Japanese and
Southeast Asians, Indonesians, the people inhabiting the islands of
Oceania, and the American natives).

We can, of course, make more subtle distinctions and create groups
inside these major groups, but as soon as we increase the number of
ethnic groups, we lose in sharpness of definition. In other words, with
a large number of groups it is more difficult to assign an individual,
with a low probability of error in classification, to any particular sub-
group. Even if we restrict our consideration to the three major groups
just mentioned, there are populations that are difficult to place. For
instance, Ethiopians are approximately half Caucasian and half African. The whole of North Africa shows an almost continuous black-
white gradient going from the southern to the Mediterranean end.
Asiatic Indians are Caucasian, but towards the east there is an increasing admixture with Orientals. Smaller groups exist in the United
States and elsewhere where the three major ethnic groups once separated have recently fused ("triracial" isolates). They are usually

isolated from neighboring populations, but only for social reasons.

What is the origin of the major ethnic groups (let us call them races)? The differences are clearly due to relative isolation in different environments. Isolation gives an opportunity for drift to operate to create differences, and the differences in environment give an opportunity for natural selection to operate, permitting adaptations to local conditions. Certainly the most conspicuous difference existing between the races is that of skin color. We know that the differentiation of black and white is due to at least four genes, although we have not identified these genes individually. There are also some differences in body build and in facial traits among the various racial groups. These all help us to recognize the racial group that an individual belongs to from his external appearance.

Some of these differences may have adaptive significance, although it is not always easy to test the various hypotheses that have been put forward to explain the differences. When we look at those genes whose inheritance is well known, we find that usually races do differ in the frequency of various alleles, as we have seen for Rh. Races tend to differ less for most other genes, such as ABO and HLA. If a random sample of genes is taken, the difference between races is, on the average, small. For most of the genes we have mentioned we do not know why they show racial differences, but they seem to be of minor importance from the point of view of survival. If some of these genes were really selectively neutral, such differences could then be said to be trivial. On the other hand, it is extremely likely that a difference such as skin color does represent a real adaptation to environmental differences. But even for skin color, the exact nature of the adaptation is not known, and only some hypotheses exist which have not been fully proved. However, the correlation between blackness of skin and intensity of solar exposure in the environment of origin is fairly obvious. What is not known precisely is how the adaptation has worked in practice. Recently the hypothesis has been put forward that white people became white because otherwise they would not be able to absorb enough ultraviolet rays in the temperate and circumpolar regions in which they live to form sufficient vitamin D. This is a rather attractive hypothesis, but it is perhaps too early to accept it fully. A careful study of this and other examples would show that it is often very difficult to explain in full detail how natural selection has worked in determining certain environmental adaptations. The only cases that are almost fully explained are those connected with malaria: hereditary anemias and G6PD deficiency.

116   The differences between races may thus be partly due to adaptation to different environments (climate, parasites, resources, etc.), and partly to drift, that is, chance. For drift to have been effective, however, some degree of isolation is important. Isolation of groups of the same species favors drift because then the groups can undergo independent random fluctuations. However, if migration between groups takes place, that is, if they exchange at least some individuals, the effect of drift in creating differences between the groups decreases—the more so, the more migration there is.

Much attention has centered on behavioral differences between races, but these are the characters for which it is most difficult to distinguish the effect of environment and of sociological inheritance from that of true biological inheritance.

We have already discussed in Section 6.5 the heritability of IQ, and the impact it may have on the origin of the difference in average IQ between U.S. whites and blacks, which is about 15 points. It should be reemphasized that the heritability of IQ, whether measured in whites or in blacks, is basically irrelevant to the problem of whether this IQ difference is environmental or genetic. There exists sufficient evidence of environmental differences, of separation between blacks and whites in the U.S., and of the plasticity of IQ towards environmental conditions, to make us think that the IQ difference might be environmental. The proof would require studying one sample of white and one sample of black babies reared partly in white and partly in black conditions. One wonders whether adoption data will ever be adequate, given the many restrictions to which adopted babies are subjected (on numbers, selection by placing agencies, and other environmental factors) to give a complete answer to the question. However, some data on transracial adoptions have now become available and seem to indicate that no important genetic difference is present.

PROBLEMS

1. Why do most genetic diseases have a low incidence?
2. How are new genes formed?
3. Random genetic drift (delete wrong alternatives):
   (a) is a change in gene frequency due to a selection difference (*true/false*).
   (b) is a change in gene frequency due to random sampling (*true/false*).
   (c) is greater, the *greater/smaller* is the size of a population.
   (d) is greater, the *greater/smaller* is migration between existing populations.
   (e) *is/is not* the only cause of differences between races.

4. *Some racial traits express adaptation to different climates.* Give at least one example for this statement and explain.

Chapter eight

The distribution of the gene for hemoglobin S (sickle cell) in Africa. Heavier shading corresponds to a higher gene frequency. There is a good correlation between the frequency of the gene and the incidence of malaria. (After Bodmer, W.F., and Cavalli-Sforza, L.L.: Genetics, evolution, and man, San Francisco, 1976, W.H. Freeman and Company Publishers.)

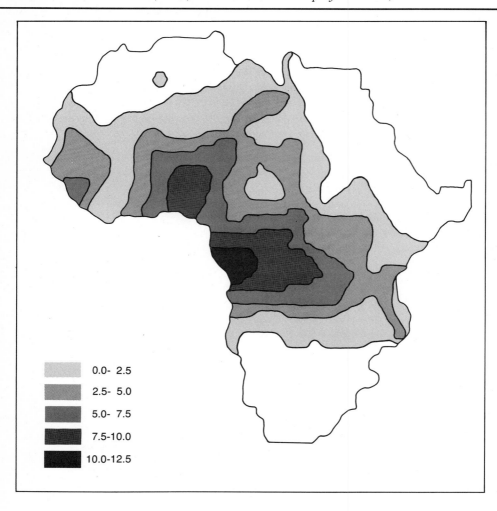

	0.0- 2.5
	2.5- 5.0
	5.0- 7.5
	7.5-10.0
	10.0-12.5

Inbreeding Depression and Heterosis

*The mating between close relatives may be dangerous
for their progeny, which has a higher than normal chance
of being homozygous. Homozygotes tend to fare badly
and, conversely, heterozygotes may flourish.*

8.1 Inbreeding and Hybrid Vigor

Animal and plant breeders have had long experience in developing
what are called *pure lines*. These consist of individuals that are as similar to one another as is practically possible. Naturally, breeders are
not interested in purity for its own sake, but in purity for those characters that they consider desirable. Pure lines are obtained by repeated
inbreeding, that is, by mating of close relatives at every generation.
The closer the relatives that are mated, the more efficient is the inbreeding procedure. It is possible to compute the probability of obtaining
genetically identical individuals, given the system of mating employed
and the number of generations. It is found that even after 20 or more
generations of very close inbreeding, some heterogeneity is still present. As we shall see, there are good reasons why heterozygosity is
very difficult to eliminate entirely.

119

120   A phenomenon commonly encountered in inbreeding observations is known as *inbreeding depression*. Inbred individuals tend to lose fertility, resistance to disease, and occasionally also size. Pure lines can be very difficult to maintain, as breeders know. However, when two independent pure lines are crossed, the reverse effects occur. The hybrids of the first generation usually show great fertility, better resistance to disease, larger size, and so on. This phenomenon is known as "hybrid vigor" or *heterosis*, and has been given very much attention by breeders in the last 50 years. In fact, it has radically changed agricultural practice and it is now being introduced into the breeding of domestic animals.

In practically all human societies an incest taboo has developed, which prevents (or at least makes extremely rare) mating between very close relatives. The origins of the incest taboo are obscure. It is not clear that recognition of the biological consequences of inbreeding may have played a role. Moreover, among animals as well, when they are given a choice, there is some tendency to avoid sex with closest relatives. On the other hand, there are social advantages to be gained from having a large circle of marriageable people. It is possible that the incest taboo developed, in part at least, to prevent the circle of marriageable people from becoming too limited. In practice, therefore, human societies are very little inbred, and there are only a few in which marriages between not-so-close relatives are encouraged. In one region in India, for instance, marriages between uncle and niece are greatly in favor; and in Japan, until relatively recently, marriages between first cousins were also popular (but this custom is now dying out).

8.2 Effect on the Progeny of a Consanguineous Marriage

Consanguineous individuals are those who have at least one common ancestor a few generations back. Thus first cousins, who are the progeny of full sibs (individuals A and B in the pedigree of Fig. 8.1), have two grandparents in common. It may happen that one of the common ancestors, say E in the figure, was heterozygous for a deleterious rare gene, such as phenylketonuria. Because this gene is rare, we can assume that all other individuals of the same generation did not have the gene. Because the gene is recessive, individual E is phenotypically normal. His children, C and D, can both receive the gene and, being

Figure 8.1
A pedigree of an individual, I, who is the progeny of first cousins, A and B. The dark half in each circle indicates a recessive gene transmitted from common ancestor E to both A and B. Since A and B are both heterozygous, individual I can be homozygous for the recessive gene which was carried in single copy by the common ancestor.

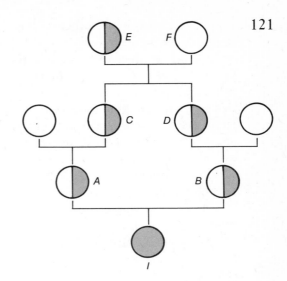

heterozygous for it, still be normal. This will occur with a given probability, which can be easily computed. In the same way, individual A can receive from his parent C, and individual B can receive from her parent D, the same deleterious gene that was originally present in E. Note that A, B, C, D, and E are all phenotypically normal, since we have assumed that this recessive gene is rare. But if A and B marry, their progeny I comes from two individuals heterozygous for the same deleterious gene. We therefore expect the progeny to show phenylketonuria with a probability of 1/4. It is easy to compute exactly the probability that a gene present in one of the common ancestors will become homozygous in the progeny of consanguineous matings; this probability is called the *coefficient of inbreeding*. In general, one can say that among the progeny of consanguineous matings, homozygotes are more frequent than would be expected in the general population from the Hardy-Weinberg rule. Therefore, we expect a higher frequency of recessive homozygote traits among the progeny of consanguineous matings. Since a fraction of these recessive genes is deleterious, we may expect that the progeny of consanguineous matings will be more susceptible to genetic diseases of a recessive nature than the progeny of random matings. It is also intuitively clear that the chance of recessive disease will be greater, the closer the consanguinity.

In most human populations, however, the deleterious effects of consanguinity are not very great. If the percentage of early deaths

122   among the progeny of a mating is used as an estimate of the total risk, it is found that first-cousin matings have about twice as high a risk as do random matings. In uncle-niece matings, the closest degree of consanguinity tolerated in marriages (and then only in certain societies), the risk is four times as great. But if we consider single recessive diseases that are very rare in the population, we find the increase of these diseases in the progeny of consanguineous matings particularly striking. It is, therefore, not surprising that when we study homozygotes for rare recessive traits (such as albinism, phenylketonuria, and many others), we usually find that there is much more consanguinity among the parents of such individuals than in the rest of the population. In fact, this is a good criterion to test whether a condition is determined by a recessive gene. A doctor who comes across a rare disease which he suspects may be inherited should question the patient to determine whether his parents are consanguineous. If they are, the disease in question may well be due to a deleterious recessive gene.

An analysis of anthropometric traits (height, weight, etc.), which are all more or less related to general size, has been made in the case of consanguineous matings in man, and minor effects have been found. Inbreeding depression in man is modest. On the other hand, because close inbreeding very rarely continues over several generations, severe inbreeding depression should not be expected in human communities except in extremely rare cases. Moreover, the study of effects of consanguinity on traits (such as anthropometric and behavioral ones) that are highly susceptible to socioeconomic conditions is complicated by the fact that consanguineous matings do not occur at random in all socioeconomic strata. In Europe consanguineous matings are especially frequent in rural societies (and also among royalty). In Japan the concentration of consanguineous matings in the various social strata is different in the rural and in the urban environments. Thus the analysis is difficult, and it may leave doubts, in some investigations at least, as to whether the socioeconomic effects have been properly considered.

8.3 The Advantage of Heterozygotes

Because of the structure of human societies, pronounced inbreeding depression is rare, and therefore cases of hybrid vigor cannot be very

striking. However, in some cases heterozygotes do have a clear-cut 123
advantage. The cases that are clear-cut all involve genes that have to
do with resistance to malaria, such as sickle-cell anemia and thalassemia, and perhaps also G6PD deficiency.

Sickle-cell anemia is a serious disease, widespread in Africa and
among people of African origin and in parts of the Mediterranean and
in south Asia, and is estimated to have a toll in the human species of
almost 100,000 deaths a year. It leads, under ordinary conditions,
to a low probability of having descendants. In Africa this probability
can be estimated at about 20% of the normal. Thus we can say that
individuals homozygous for hemoglobin S (and therefore affected by
the severe form of sickle-cell anemia) have a fitness of 20%. Heterozygotes, on the other hand, are practically normal individuals who are
subject to danger only in special situations (for instance, if the oxygen
content of the air is lowered considerably, as might happen at very
high altitudes or in a plane that loses its internal pressure). In general,
heterozygotes can be distinguished from "normal" individuals (having
no hemoglobin S, i.e., homozygous for the more common gene, making hemoglobin A) only by laboratory tests. Their fitness in a normal
environment is close to 100%. However, in an environment where
malaria is prevalent the situation changes radically. It is clear from
various sources of evidence that normal individuals have less defense
against the malarial parasite than do heterozygotes. Thus, on the one
hand, we expect natural selection to eliminate the hemoglobin S gene
because it behaves as an almost fully lethal recessive, in that most of
the homozygotes for it die. On the other hand, in the presence of
malaria the frequency of deaths among homozygotes for hemoglobin
A is somewhat higher than that among heterozygotes (the fitness of
homozygote normals in a malarial environment is 80 to 90% that of
heterozygotes). If only the heterozygotes survived, there would be only
SA individuals among the adults, and thus in each adult half the genes
would be *S* and half would be normal (*A*). This is not the case, however, and in the presence of malaria homozygous normals fare better
than do individuals homozygous for the *S* gene. This is why *S* gene
frequencies tend to remain high but less than 50%. In malarial areas
20 to 30% of the population usually consists of heterozygotes.

If malaria is eradicated in a population with a high frequency of
S genes, then the heterozygote advantage disappears and the *S* gene
becomes a deleterious recessive. Natural selection against lethal reces-

124 sives is slow, and therefore the process of elimination of *S* takes a number of generations. Thus it is not surprising that present data indicate that black Americans, who have been living in an environment free of malaria for some generations, do have a somewhat lower frequency of the *S* gene than their ancestors probably had when they were first brought to America. In Central and West Africa, where malaria is still prevalent, there is no reason for the gene to disappear. In fact, there it is still advantageous for an individual to have one gene for hemoglobin S.

A very similar but more extreme case is that of thalassemia. This is a recessive lethal, and the probability that a thalassemic homozygote will reach maturity is practically zero. The gene can be considered recessive in the sense that under normal conditions heterozygotes are practically indistinguishable from normals except by laboratory tests. Again, if there were no special reason for the heterozygote to be at an advantage over the normal, this gene should have been eliminated or reduced to very low frequencies. Indeed, its frequency is very low in almost all populations that have not been exposed to malaria. In some malarial areas, however, such as many parts of South Asia, Greece, Sardinia, and a region around Ferrara in northern Italy, up to 20% of the population may be heterozygotes. Here again the heterozygote is clearly at an advantage over the homozygote normal when there is a high chance of malarial infection.

8.4 Polymorphisms

Polymorphic genes are those for which there exists more than one allele in a population and for which all alleles are substantially frequent (i.e., their frequencies are substantially different from 0 or 100%). Thus ABO is a polymorphism, Rh is another, and so is HLA. We have mentioned that an analysis by immunological and by electrophoretic techniques has revealed that a large proportion of genes is polymorphic. But the alleles of one gene may not differ much from one another in terms of their effect on the individual's capacity to prosper in a given environment, that is, in terms of their fitness. Some people maintain that many polymorphisms are selectively neutral. The reason why so many allelic forms of genes exist may be that mutation has introduced them, and chance, in the form of random genetic drift, keeps them oscillating at frequencies that change slowly and

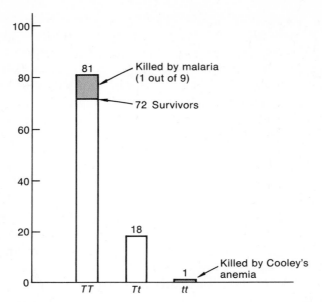

Figure 8.2
Here we illustrate genetic equilibrium for the thalassemia gene in the presence of malaria due to heterozygous advantage. At birth, with a gene frequency of 10% for t, *there are by Hardy-Weinberg rule, 81% of* TT, *18% of* Tt, *and 1% of* tt *individuals (represented by bars). Before they become adults, malaria kills one out of nine* TT *individuals (black section of the bar). Cooley's anemia is the hereditary disease killing practically all* tt *homozygotes for the thalassemia gene. Out of 100 newborn, among adults there remain only 72 homozygous normals and 18 heterozygotes. This now totals to 72 + 18 = 90, and percentages have to be readjusted to this new total of 90 survivors. The gene frequency of* t *is still 1/2 (18/90) = 10%, as before. In the next generation there will again be found 81% TT, 18% Tt, and 1% tt. The situation is at equilibrium, that is, the gene frequency does not change as long as malaria kills one out of nine TT individuals, and Cooley's anemia all tt homozygous individuals.*

randomly. But some of these alleles must have a definite selective advantage, even if small, and are slowly but surely gaining prevalence in the population. (See Chapter 7.)

If we could look at the same population at two widely separated points in time (thousands of generations), we might find radical changes in the frequencies of various alleles. Many alleles might have disap-

126   peared and been substituted by others either through chance or because those that took their place gave their carriers a better adaptation to the environment. But genes in which the heterozygote has an advantage over both homozygotes (such as the genes for sickle-cell anemia and thalassemia) give rise to a peculiar kind of polymorphism, which is called *stable*, or *balanced*, because gene frequencies tend to reach a certain value and return to it if disturbed (for example, by chance fluctuations). In other words, both alleles will remain present, and unless the environment or other conditions change, the numerical composition of the population in terms of those genes will tend to remain the same.

We show in Fig. 8-2 how the equilibrium for a balanced polymorphism works in the case (for instance) of thalassemia. Let us take a population in which 20% of the adults are heterozygotes and 80% are homozygotes for the normal gene. All the homozygotes for the thalassemia gene have died previously as a result of serious anemia. The gene frequency of the thalassemia gene is easy to compute; it is half the frequency of heterozygotes, and therefore 10% or 0.1. Applying the Hardy-Weinberg formula, we expect that among all children born there will be $(0.9)^2 = 0.81$ or 81% homozygotes for the normal gene (TT), 18% heterozygotes (Tt), and 1% homozygotes for the deleterious gene (tt). The latter will all die before reaching the age of reproduction. We can ask: "Will these frequencies remain unaltered in the next generation, as we should expect if this polymorphism is really stable?" They will, if the strengths of selection against the two homozygotes are in certain ratios, which can easily be computed. We shall not give the details of these computations, but simply show their results (see also Fig. 8.2).

Let us assume that there is malaria and that all heterozygotes survive (this is an oversimplification). If we assume that exactly one out of nine homozygote normals die of malaria, so that exactly 8/9 survive, the initial 81% of homozygote normals is now reduced to 72% (= 8/9 of 81). Thus those who survive to adulthood and can therefore reproduce are in a ratio of 72 homozygous normals to 18 heterozygotes (all the homozygous thalassemics have died). Note that in this new generation we still have the same ratio of TT to Tt among adults as we started with, 80 to 20, and therefore the gene frequency has not changed. Even if perturbations were introduced (by mutation, drift, or any other accidental changes in gene frequencies), this equilibrium due to heterozygote advantage would be restored by selection. In other words, the

gene frequencies would tend to return to the original value, which is   127
determined by the ratio of the fitnesses of the various genotypes.

In the numerical example just given, we chose the fitness of the homozygote normal as 8/9 that of the heterozygote so as to give equilibrium at 80% homozygote normals and 20% heterozygotes among adults; this procedure may seem artificial. The strength of selection may indeed vary with time and place, but it seems that the equilibrium for thalassemia is relatively ancient. The same can be said of sickle-cell anemia, for which it can be shown that gene frequencies will have stabilized because of heterozygote advantage, and fitnesses computed directly agree with those computed from gene frequencies on the assumption that the system has reached equilibrium. If malaria is eradicated, then the equilibrium due to heterozygote advantage will be destroyed, and both sickle-cell anemia and thalassemia will slowly disappear over a very large number of generations. Here is one case in which improvement in medicine and hygiene (i.e., control of malaria) also automatically eliminates a genetic disease. In other cases we may fear the reverse. We have learned how to control diabetes, a disease of sugar metabolism caused by a (probably genetic) deficiency of production of insulin. We can therefore save the lives of diabetics by administering insulin to them (and by other techniques). We have thus reduced the action of natural selection that kept diabetes relatively rare, and can therefore expect an increase in the number of diabetics in future generations. But again, this change will be very slow. The same can be said of hemophilia, galactosemia, phenylketonuria, and all other genetic diseases that are treated by substitutional therapy or special diets.

PROBLEMS

1. Why does the progeny of consanguineous marriages show a higher incidence of genetic disease, and which diseases does it show in particular?

2. Malaria is a powerful selection agent. What examples of genetic adaptation to malaria can you offer?

3. A genetic polymorphism (delete wrong answer):
    (a) is the presence of a rare variant in a population (*yes/no*).
    (b) is always balanced (*yes/no*).
    (c) may change over time (*yes/no*).

4. What conditions will cause the gene frequency of a balanced polymorphism to change?

Chapter nine

As a prelude to genetic engineering, our knowledge of the structure of chromosomes of both lower and higher organisms and of viruses interacting with them is deepening. Here are chromatin-like forms of Simian Virus 40 extracted from infected mammalian cells. Three nucleoprotein complexes are shown: each appears as a series of about 20 globular particles (called nucleosomes) interconnected by thin filaments. Each nucleosome contains cellular proteins and SV40 DNA condensed fivefold in length. The internucleosomal filaments are made of DNA free of proteins. Bar shown is 200 mm long. (Courtesy C. Cremisi, P.F. Pignatti, O. Croissant, and M. Yaniv.)

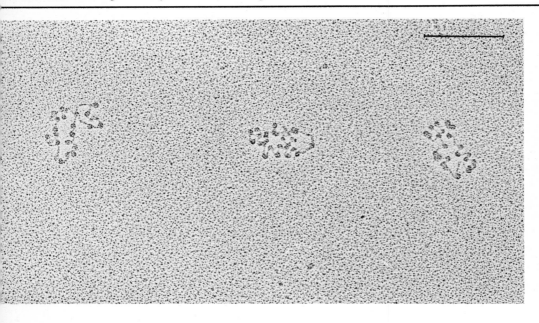

The Future of
Human Genetics

*It is not easy to improve mankind, but molecular
biology may offer better methods than those suggested by
the experience of animal and plant breeders.*

9.1 Genetic Engineering

Few people would dispute that improving our health is one of the
major aims to be encouraged. Developing knowledge shows that many
diseases—many more than we suspected—have a genetic basis. Some
of them have come under medical control; even if we do not yet know
how to correct the genes themselves, we can in some cases correct
their wrongdoings by medical means. Insulin therapy for diabetes
is an early example of this type of correction. Many other hormonal
diseases are treated in a similar way. So also is hemophilia. Therapy
that consists in giving the organism the product that it cannot manufacture for itself because of a defective gene is called substitutional
treatment. The major problem with this kind of therapy is that it may

132   have to be lifelong, with frequently repeated administrations of the needed substance. If we could introduce a normal gene into the cells of the patients and make sure that it keeps active, we would no longer need to treat individuals continuously over a long period of time.

It is very encouraging to note that the recent progress in molecular genetics gives reasonable hopes of performing surgery on defective genes and chromosomes, thus giving a normal life to individuals who would otherwise be afflicted with some kind of genetic disease. How is it possible to carry out this kind of surgery at the ultramicroscopic level on a structure like DNA, a thread 2 millionths of a millimeter thick? The hope comes from experiments on bacteria, and on bacterial and animal viruses. Enzymes have been discovered, called restriction enzymes, which can cut DNA at certain points, that are specified by the local sequence of nucleotides. The cutting may take place, for example, in such a way as to generate, from a double-stranded DNA segment,

$$\downarrow$$
$$\ldots\ldots\ldots \mathrm{ATCCACCTACCG}$$
$$\ldots\ldots\ldots \mathrm{TAGGTGGATGGC}$$
$$\uparrow$$

two fragments by breaks at different points of the two strands. For instance, by breaks at the arrows, the fragments formed are

$$\ldots\ldots \mathrm{ATCCAC} \qquad \mathrm{CTACCG}\ldots\ldots$$
$$\ldots\ldots \mathrm{TA} \qquad \mathrm{GGTGGATGGC}\ldots\ldots$$

The endings (which are always the same for a given enzyme) are complementary. They therefore tend to stick together; another enzyme can repair the breaks and reconstitute the original DNA. If on the other hand one treats with the restriction enzyme a mixture of DNA from, say, a bacterium and an insect like Drosophila, then both DNA's will break in the same way, and the repair enzyme, applied to the mixture, may stitch together—instead of the fragments of bacterial DNA—a fragment of DNA from the bacterium and one from Drosophila. This "hybrid" DNA (a sort of longitudinal hybrid) can be transferred into a new bacterium and it can enter the bacterial chromosome. This experiment has now been carried out a number of times, and the extraneous piece of DNA has been shown to function, at least partially, in its new host. The most immediate applications are those using bacteria to manufacture proteins for human use, like hormones (especially "polypeptidic" ones, like insulin and other small proteins, which can be synthesized in the laboratory only with the greatest difficulty and could be produced much more economically by genetic engineering), enzymes, and perhaps even antibodies. These products could thus be made in large quantities; many of them already have important clinical applications, and others could have industrial or agricultural applications. Therefore a true technological revolution is at hand, and the first results may come fairly soon.

A more ambitious task is to carry out this DNA "stitching" inside the cells of a diseased individual. In principle, this procedure too is possible, but it must make use of vehicles for bringing the "right" DNA inside the cell. The vehicles which one can think of today are viruses, which may not be devoid of risk. To make the procedure efficient and harmless will undoubtedly require much more work. The research itself may involve dangers which have been anticipated and are being considered with great care. Mostly, these dangers consist of generating new bacteria or viruses endowed with new unpleasant properties. Fear of these dangers has prompted the research workers involved in the work to prescribe a considerable number of precautions which should defuse these—fortunately hypothetical—biological bombs.

Even if the applications of the new technique of "genetic engineering" (genetic surgery would be also an appropriate name) to genetic disease are still fairly remote, it can be hoped that the first results may be forthcoming in one decade or in a few decades.

9.2 Eugenics

Most treatments for genetic disease are necessarily euphenic in character; that is, they can correct the phenotype of the individual, but his descendants will still inherit the defective gene. A movement called *eugenics*, which has become popular since the end of the last century, claims that the human species can be improved by the application of the methods of artificial selection used by breeders of domesticates. That would mean weeding out the "bad" genes and giving higher chances of reproduction to "good" genes. The former is called negative, and the latter positive, eugenics. Positive eugenics suffers from

134   some serious drawbacks. One of these is that for many genes it is difficult to say whether they are good or bad. Some genes, in fact, have been shown to be good in one environment and bad in another, and we do not have enough control of the environment to commit ourselves to selection for one type only. Furthermore, for most traits that are of social and behavioral importance, we really do not know whether they are genetically or environmentally determined. Finally, genetic variety is essential in many ways. Some loss in genetic variety would be an inevitable by-product of programs of positive eugenics, and would certainly be extremely dangerous at our stage of knowledge. Even in recent times some people have strongly advocated the practice of positive eugenics, for instance, by the conservation of the sperm of outstanding men by techniques already used for bull semen, and by the promotion of artificial fertilization of women with selected semen. Apart from other obvious difficulties, it would certainly be difficult to reach an agreement on whose sperm should be propagated. Should we choose, for instance, Mao's sperm, Nixon's sperm?

It is not difficult to recognize that these ideas all derive from the very successful practices of animal and plant breeders. The domestication of plants and animals, which goes back to more than ten thousand years ago in some areas, has been constantly accompanied by voluntary or involuntary genetic manipulation of the domesticates. In this century the pace of genetic improvement of plants and animals has greatly increased, thanks to the introduction of sound scientific methods. However, even though relatively fast, the improvement of stocks by selective breeding is still a slow process. More importantly, selective breeding in humans would demand an external control of the reproductive process or at least a very heavy interference with it. With very few exceptions, this is incompatible with our most deeply rooted customs and drives. It also lends itself easily to serious misinterpretation, as well as to political exploitation ranging from racism to genocide. Like every technological development, genetics too can be put to good or bad use. The choice is ours.

Negative eugenics, or the selective outbreeding of genetic disease, is somewhat less objectionable than the positive approach, but suffers from other drawbacks. A type of negative eugenics practiced in some countries is the sterilization of individuals who carry serious genetic diseases. However, computations show that this method, apart from all humanitarian considerations, is not very effective in reducing the incidence of defects in the future generations. In fact, the benefit that can be expected is usually small or negligible.

In general, the eugenics movement, which still counts a certain number of supporters in its ranks, suffers from the drawback that it can only promise very slow results; sometimes eugenic programs can be totally ineffective or potentially dangerous. In general, it is difficult to become very excited about prospects that are ten or a hundred generations removed. The tempo of cultural change makes improvement desirable or necessary within a much shorter term. It is encouraging that genetic engineering can offer a very valid alternative. Note, however, that the results of genetic engineering do not, or do not necessarily, affect the progeny of an individual, but may very well be limited to the phenotype of the patient. One can anticipate them to be usually euphenic, not eugenic.

9.3 Genetic Counseling and Prophylaxis of Genetic Disease

A recent development in medicine is that of genetic counseling. Couples who have had a child affected by some congenital malformation or genetic disease—or who, for one reason or another, fear to have such a child—seek advice on what the chances are that their future progeny may be affected by some such disease or malformation. Advice can often be given only in terms of probability; recognized heterozygotes for cystic fibrosis, for instance, have a chance of 1/4 of having an affected child; an individual affected by Huntington's chorea has a 50% chance of having a child thus affected, and so on. For a certain number of diseases, the help can go further: a fetus can be diagnosed for a certain number of genetic diseases in time for elective abortion. The process takes the name of *amniocentesis* (see Fig. 9.1). By this process, all chromosome aberrations and many inborn errors of metabolism can be selectively eliminated. These procedures are carried out in specialized centers of medical genetics, which by now exist in a number of cities. Monitoring of diseases whose effects can be avoided by proper dietetic treatment, such as PKU or galactosemia, is also another type of genetic prophylaxis which is now being extended to more countries and more diseases.

These practices should not be confused with negative eugenics, which aims at the population, rather than at individual welfare. In fact, they can be slightly dysgenic, that is, they can serve to increase rather than decrease the incidence of a given genetic trait or malformation in later generations. The dysgenic effect is, however, negligible in practice.

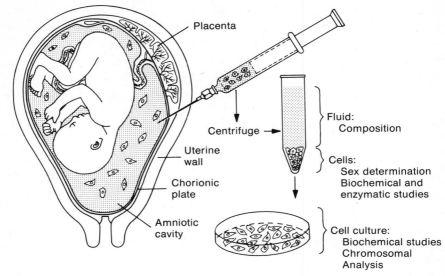

Figure 9.1
Amniocentesis is a technique for prenatal diagnosis that is carried out by inserting a sterile needle into the amniotic cavity and withdrawing a small amount of fluid (see text). An amniotic tap at 16 weeks of gestation provides ample time for diagnostic workup and, if indicated, a therapeutic abortion. (From Hood, L. E., Wilson, J. H., and Wood, W. B.: Molecular biology of eucaryotic cells, *Menlo Park, California, 1976, W. A. Benjamin, Inc.)*

9.4 Genetic Variety

An important fact which must be kept in mind is that all human individuals are different from one another. Many of these differences are subtle: some may be of no significance whatsoever, but some do matter. This variety is not at all undesirable. In fact, the complexity of our social systems necessitates a great variety of individuals. Let us take an extreme (and somewhat hypothetical) example. When we spoke about schizophrenia, we indicated that there is a great number of slightly schizophrenic individuals who are virtually normal, with only a trace of the symptoms that occur in a major case of the disease. These personalities are sometimes referred to as schizoids. There is also some evidence that schizoid types are more frequent among the relatives of schizophrenics, indicating that there may be a biological background. If we wanted to be more precise, we could measure the degree of "schizophrenicity" and develop a "schizophrenia quotient" (SQ) along the lines of an intelligence quotient (IQ). If we did that, we might find that most good artists have a high SQ and most good administrators have a low SQ. If we had only artists and no administrators, we would probably very soon be in a state of economic chaos, but if we had only administrators and no artists, we might have to give up movies, theaters, paintings, and all sorts of art and literature, in other words, a great part of our culture and entertainment.

There are many other dimensions in which society needs the coexistence of a great number of different types. Our knowledge of the inheritance of socially important traits, such as criminality or drug addiction, is still extremely vague or nonexistent. But when we come to study behavioral traits, two rules seem fairly general: (1) genetics is never all-important, and environmental effects frequently play a very important role; (2) the separation of the effects of genetics and of environment is difficult, and the conclusions reached are usually uncertain. Further study of the underlying genetic phenomena at the biochemical and physiological level is the only way to improve our knowledge. Meanwhile, hasty conclusions should be avoided.

9.5 Some Other Problems

A possible genetic technique is the separation of sperm containing X and sperm containing Y so as to produce babies of the desired sex. This would unquestionably be wonderful for cattle breeders. In man, however, there has been disagreement on whether such a technique, if and when it becomes feasible, would be socially useful. I prefer to be an optimist and assume that it would contribute to the solution of one of the major problems in the world today, birth control, because in some families at least, the quest for a child of the desired sex has led to large and not always successful increases in family size.

In general, we can expect an increasing contribution from the application of the techniques of biochemical genetics, which can tell us about the metabolic disturbances accompanying disease. This knowledge will give us a basis for diagnosis and therapy of diseases that are at least partially of genetic origin. We can also expect that genetic maps will become more detailed, and this additional knowledge will serve the purpose of increasing the power of genetic prediction.

138   A big problem for human genetics, and for genetics of higher organisms in general, is understanding the mechanisms of gene action, especially in differentiation. We know that all cells have the same genes, but in each cell some genes are active and others are switched off. We know practically nothing of how such phenomena take place. We are, however, convinced that the whole of the development of an organism, and much of its functioning, is due to regulatory mechanisms, probably genes which produce regulatory substances that switch other genes on and off. Some of these substances are known; they are hormones (e.g., sex hormones, and hormones produced by the pituitary gland, the thyroid gland, etc.). Others may act by a mechanism similar to those already found in bacteria for switching enzyme production on and off according to needs. It is possible that some diseases, especially dominant ones, may be due to mutation of these regulatory genes. This is another field in which we expect much progress in the next decade or two.

There may be much to learn about innate differences in response to various educational methods. It is unfortunate, perhaps, that in this field it is most difficult to distinguish between economic, social, and genetic factors. In short, human genetics has taught us that all are far from equal; but also that many of the differences are superficial, and very many of them are probably of very little importance. But there are also differences that are important, and many of these involve some kind of disease.

In recent years technology has offered progress too rapidly for the social structure of the population to be able to absorb it. Severe social imbalance has resulted. Excessive population growth and perhaps some mental diseases and increasing criminality are among the consequences. Freedom of reproduction may have to be limited for nongenetic reasons if education does not succeed. Our population is growing too rapidly, and we must impose rigid birth control practices on ourselves. But this should be done by education rather than coercion, if possible, and along with education, genetic counseling might also be usefully introduced. It should be remembered that in the case of recessive genetic diseases (including sickle-cell anemia) genetic counseling does *not* imply that heterozygotes should not marry, or should have no children, but simply that they should not marry other heterozygotes for the same gene. This is not a serious limitation for heterozygotes. We are not yet fully prepared socially and technically to apply this principle on a large scale, but there is sufficient scope and interest for genetic counseling so that many centers are already in existence.

In summary, human genetics is a rapidly developing field, from which we can hope to obtain important contributions to our welfare.

Solutions to Problems

Chapter One

1. (a) mitosis (b) meiosis (c) meiosis (d) mitosis (e) meiosis

2. (a) male (b) female (c) male (d) The presence of at least one Y is necessary and sufficient for an individual to be male.

3. (a) 45 chromosomes with a translocation of a 21 chromosome to another autosome (b) of the order of 1/4 to 1/3 (in some special cases higher or lower, but these are not discussed in the text).

4. An increased risk of Down's syndrome.

Chapter Two

1. the male sex

2. (a) no (b) no (c) yes, 1/2 (d) no (e) yes, 1/2 of them

3. (a) no (b) yes, all (c) both sons and daughters

4. (a) In the sperm (It must have an XY sperm and the mother must have given only one X.). (b) We cannot say if the event happened in sperm or egg. (c) In the egg, as the father cannot contribute an X with the color-blindness gene.

5. Since only half of the X chromosomes of the female are active, we expect an *HH* female, and an *H* male, to produce the same amount of *H* globulin. (This phenomenon is called dosage compensation.) An *Hh* female produces half the amount of globulin compared with an *HH* or an *H* male).

142   Chapter Three

1. (a) both heterozygotes   (b) 1/4
2. no
3. (a) yes   (b) no
4. half taster and half non-taster
5. all blind if it is the same recessive for which both parents are homozygous; all normal otherwise
6. sickle cell anemia Rh−: 1/16; sickle cell anemia Rh+: 3/16; non-anemics Rh−: 3/16; non-anemics Rh+: 9/16.

Chapter Four

1. (a) frequency of recessives (q) = $\sqrt{0.3}$ = 55%, of dominants (p) = 45%   (b) $p^2/(p^2 \times 2pq)$ = .2025/(.2025 × .4950) = 29%
2. 50%
3. 3.7%
4. 1/400 million
5. (Exp 1) expected 6.25, 37.5, 56.25; χ^2 = 0.71
   (Exp 2) expected 5.06, 34.88, 60.06; χ^2 = 0.29
   (Exp 3) expected 6.50, 38.0, 55.50; χ^2 = 1.73
   (Exp 4) expected 7.29, 39.42, 53.29; χ^2 = 0.75

Chapter Five

1. (a) yes   (b) no   (c) no   (d) yes   (e) no
2. 36%
3. 43%
4. (a), (b), (e)

Chapter Six

1. 25%

2. (a) 100% (b) There are environmental factors affecting the trait (but there might also be genetic ones). (c) It strengthens somewhat the hypothesis of a genetic component. (d) adoption data

3. adoption data

Chapter Seven

1. Because they are mostly maintained by mutation against adverse selection, and mutation is usually low.

2. by duplication

3. (a) false (b) true (c) smaller (d) smaller (e) is not

4. skin color, body size, body shape

Chapter Eight

1. Recessive diseases are increased in frequency among the progeny of consanguineous matings because the relative frequency of homozygotes is increased.

2. sickle-cell anemia, thalassemia, G6PD

3. (a) no (b) no (c) yes

4. A change in environmental conditions determining changes of the fitnesses of homozygotes relative to heterozygotes. Also, drift.

References

Bodmer, W., and Cavalli-Sforza, L., 1976. *Genetics, evolution and man*. San Francisco: W. H. Freeman and Co. (a textbook of human genetics and evolution).

Cavalli-Sforza, L., and Bodmer, W., 1971. *The genetics of human populations*. San Francisco: W. H. Freeman and Co. (a reference book that deals more specifically with the subjects here shortly described in Chapters 4 to 9).

Harris, H., 1975. *The principles of human biochemical genetics*. Amsterdam and New York: North Holland (a very clear introduction to inborn errors of metabolism and molecular genetics in man).

Hood, L. E., Wilson, J. H., and Wood, W. B., 1975. *Molecular biology of eucaryotic cells—a problem approach*. Menlo Park: W. A. Benjamin (molecular and cellular biology of higher animals).

McKusick, Victor A., 1975. *Mendelian inheritance in man*, 4th ed. Baltimore: The Johns Hopkins University Press (a catalog of Mendelian traits known in man).

Stern, C., 1973. *Principles of human genetics*, 3rd ed. San Francisco: W. H. Freeman & Co. (a general textbook of human genetics).

Watson, J. D., 1976. *The molecular biology of the gene*. Menlo Park: W. A. Benjamin (a classical introduction to the subject).

Glossary

achondroplasia a disturbance of growth of the long bones, often resulting in a special form of dwarfism, usually dominant.

agglutination the formation of clumps —usually of red cells of the blood, held together by antibodies attached to antigens on the cells' surfaces.

alleles alternative forms of the same genetic locus.

amino acid small molecules that are the building blocks of proteins. There are 20 amino acids that commonly make up proteins. All amino acids have the same general structure with one acidic (carboxyl) and one alkaline (amino) end, but they differ in the side groups (R groups).

amniocentesis a clinical procedure by which a few milliliters of the amniotic fluid surrounding the fetus are withdrawn. The fluid and fetal cells contained in the fluid may then be subjected to tests for various genetic diseases.

anaphase the period of cell division during which the chromosomes begin

migration toward opposite poles of the cell.

aneuploidy a karyotypic abnormality resulting from the presence of extra chromosomes or absence of chromosomes, such that the karyotype is neither haploid nor an exact multiple thereof.

antibody a protein produced in the immune reaction with property of binding to specific foreign molecules.

antigen a molecule that stimulates the production of specific antibodies.

autosome any chromosome other than the sex chromosome.

backcross a cross between a heterozygote (*Aa*) and a corresponding homozygote (*AA* or *aa*).

balanced polymorphism a polymorphism that is stable (tends to remain unchanged over time) and is probably maintained by advantage of the heterozygote over both homozygotes.

Barr body (sex chromatin) a mass of chromatin in the nucleus of resting cells, resulting from inactivation of an X chromosome. A cell ordinarily contains a number of Barr bodies that is equal to the number of X chromosomes minus one.

Almost all terms appearing in this glossary are taken from the book Genetics, Evolution and Man *by W. Bodmer and L. Cavalli-Sforza published by W. H. Freeman and Company of San Francisco. Copyright © 1976, with the kind permission of the publisher and of Prof. W. Bodmer.*

147

centromere (also called **kinetochore** or **primary constriction**) the constricted portion of the chromosome by which the chromosome attaches to the spindle fibers at mitosis.

chimera an individual whose cells are not all of the same genotype. DZ twins are occasionally chimeras. See **mosaic,** which is distinct from a chimera.

chi-square (χ^2) test a statistical test used to determine if a set of observed frequencies differs (to a degree that would be improbable by chance alone) from those expected on the basis of a specific hypothesis.

chromatin the basic substance of the chromosome, including both proteins and DNA. This term essentially means "chromosome material."

chromosome a threadlike body found in the nucleus of a cell and containing the genes. Under the light microscope, the chromosomes are not visible during interphase.

chromosomal aberrations karyotypic alterations involving whole chromosomes or portions of them, sufficiently large to be detectable through light microscopy.

codominant said of two alleles that are both expressed in the heterozygote.

consanguinity two or more individuals are said to be consanguineous if they have a common recent ancestor (usually not further back than three or four generations).

coupling dominant alleles at two different loci are said to be in coupling in a heterozygote if they are located on the same chromosome. Thus, the double heterozygote *AB/ab* is said to be in coupling, while *Ab/aB* is said to be in **repulsion.**

crossing–over the process of exchange of genetic information between two homologous chromosomes, presumed to occur through breakage of both chromosomes at homologous sites followed by reunion after exchange.

cystic fibrosis a recessive genetic disease involving abnormal function of the pancreas and other secretory glands, especially common among Caucasians.

Darwinian fitness the fitness of a given genotype in a given environment is measured by its relative contribution to the ancestry of future generations— that is, by the change in the frequency of this genotype from one generation of parents to the next generation of parents. It depends on both fertility and survival.

deletion at the molecular level, the removal of one or more bases from a DNA sequence. At the cytological level, the absence of a segment of a chromosome (also known as a deficiency).

deoxyribonucleic acid (DNA) a polymer of nucleotides in which the sugar residue is deoxyribose. DNA is found primarily in the double-helical conformation.

diploid a chromosome complement that contains two copies of each chromosome. Normal human somatic cells are diploid.

dizygous (DZ) twins twins that arise from two different eggs fertilized by two different sperm. The genotypes of dizygous twins have no more related than those of any two sibs.

dominant an allele manifesting its phenotypic effect also in the heterozygotes; a trait determined by a dominant allele.

Down's syndrome (trisomy 21, or **mongolism)** a syndrome characterized by mental, behavioral, and physiological defects, caused by the presence of an extra copy of the genetic material contained on chromosome 21. The third copy of this information may be present as an extra chromosome 21, or as a segment of it translocated to another chromosome.

drift see **random genetic drift.**

duplication presence of two copies of a chromosome segment, usually in the same chromosome—sometimes in immediate sequence (tandem duplication), or elsewhere in the same or other chromosomes.

dysgenic due to, or determining, an increase in the frequency of deleterious genes; the opposite of *eugenic*. Dysgenic effects may be spontaneous, or they may result from medical or social interventions that improve the fitness of the handicapped.

egg (egg cell) the female **gamete.**

electrophoresis a technique for separating molecules, particularly proteins, according to the overall electric charge of the molecules.

epistasis nonadditive interaction between two or more different loci.

eugenic due to, or determining, a decrease in the frequency of deleterious genes.

eugenics a program of decreasing the frequency of deleterious genes in a human population (negative eugenics) or of increasing that of advantageous genes (positive eugenics) through artificial selection against the genetically handicapped or in favor of the types considered especially desirable.

enzyme a protein that catalyzes a specific chemical reaction.

fitness see **Darwinian fitness.**

galactosemia a recessive genetic disease involving incapacity to utilize the sugar galactose, a normal component of milk.

gamete the haploid cell generated by meiosis that may fuse with another appropriate gamete to form a zygote. In a bisexual species, a gamete is either male (sperm) or female (egg).

gene a segment of chromosome with a detectable function. Used as a synonym for **locus** and sometimes for **allele.**

gene frequency the proportion (of all alleles at a locus) in which a given allele is found in the individuals forming a specified population.

genetic code the code that relates nucleotide sequences in nucleic acids to amino-acid sequences. Each triplet of nucleotides designates a particular amino acid; thus, the genetic code allows the translation of information stored in DNA and the use of that information in protein synthesis.

genetic equilibrium a state reached by a population when gene and genotype frequencies do not change in successive generations.

genetic marker a gene mutation that has phenotypic effects useful for tracing the chromosome on which it is located.

genetic screening a systematic testing of individuals to ascertain potential genetic handicaps in them or in their progeny—handicaps that may require treatment or prophylaxis.

genotype the genetic constitution of an individual at one or more loci.

germ cells gametes, or their precursors.

germ line the line of cells that produce gametes.

haploid a cell is said to be haploid if it contains one copy of each chromosome.

Hardy-Weinberg law a rule for predicting genotype frequencies on the basis of gene frequencies, under the assumption of random mating in the absence of selection.

hemoglobin a globin (globular) molecule found in the blood; it transports oxygen from the lungs to other tissues. The molecule is a tetramer (formed by four polypeptide chains); in the adult, it contains two α chains and two β chains.

hemolytic anemia anemia due to excessive destruction of red cells. There are many genetic diseases in which this is the outstanding manifestation, usually recessive.

hemophilia an X-linked recessive disease in which clotting time is abnormally long.

heterosis tendency of hybrids to be "better" than either parental line from the crossing of which they originated.

heterozygote a cell or individual that is heterozygous.

heterozygous having different alleles at a given locus on homologous chromosomes.

histocompatibility capacity to accept a tissue or organ graft.

homologs chromosomes (or chromosome segments) that carry genes governing the same characteristics, that have similar morphology, and that pair during meiosis.

homozygote a cell or individual that is homozygous.

homozygous having the same allele at a given locus on homologous chromosomes.

hybrid the progeny resulting from a cross between different parental stocks.

in utero (Latin) in the womb.

in vitro (Latin, in glass); pertaining to experiments done on cells grown outside of the animal.

in vivo (Latin, in the living); pertaining to experiments on a whole living animal.

inborn errors of metabolism inherited disorders that can be explained as genetic blocks in specific metabolic pathways, usually due to recessive alleles determining a decreased activity (or the absence) of a specific enzyme.

inbreeding when consanguineous (closely related) individuals mate, their progeny is said to be inbred. Such mating is called inbreeding.

incompatibility the presence in a fetus (or graft, or donated blood) of antigens that can evoke an immune response in the mother (or graft recipient, or blood recipient).

inversion a chromosomal aberration that arises when two breaks occur in the same chromosome and the region between the breaks is reinserted after a 180° rotation.

karyotype the chromosome complement of any organism, analyzed according to size and banding patterns of each chromosome.

Klinefelter's syndrome a syndrome caused by the presence of an extra X chromosome in a male karyotype (XXY). Affected individuals are phenotypically male, but with underdeveloped gonads; other physical and behavioral problems are present in many cases.

lethal an allele that kills all carriers (or only homozygotes in the case of a recessive lethal) before reproductive age and usually in the first years of life.

linkage the presence of two or more loci on a single chromosome, causing a tendency for alleles at the linked loci to be inherited together. Linkage is observed only when the loci are sufficiently close to one another; crossing-over can lead to random assortment of loci that are far apart on the same chromosome.

locus (plural, loci) position of a gene on a chromosome. Used as a synonym for gene or cistron, but not for allele.

meiosis (reduction division) a series of two modified mitoses, generating haploid gametes from a diploid precursor cell.

metaphase the phase of mitosis or meiosis in which the condensed chromosomes attached to the spindle fibers line up on an equatorial plane between the two poles of the cell.

mitosis the process of cell division in somatic cells, in which duplication and assortment of the chromosomes ensures the identity of genetic information in the parental and the two daughter cells.

mongolism an old term (now seldom used) for **Down's syndrome.**

monosomy the presence of only one member of a chromosome pair.

monozygous (MZ) twins   twins that develop from a single zygote, which divides to give rise to two complete embryos. Monozygous twins have identical genotypes.

mosaicism   the presence in an individual of two or more cell genotypes arising by mutation or by chromosomal aberration (including nondisjunction).

multifactorial trait (polygenic trait)   a trait whose phenotypic expression is influenced by the cumulative effects of many genes.

mutagen   a physical or chemical agent that increases the mutation rate.

mutation   a heritable change in the genetic material, or its detectable effects in the phenotype.

natural selection   the process by which changes occur spontaneously in the proportions of genetic types within populations of a living organism, due to differences in the (Darwinian) fitnesses of these genetic types in the existing environment.

nondisjunction   failure of homologous chromosomes to separate during the first stage of meiosis (primary nondisjunction) or of chromatids to separate in the second division of meiosis (secondary nondisjunction). Nondisjunction can also occur in mitosis.

nucleic acids   deoxyribonucleic acid (DNA) and ribonucleic acid (RNA).

nucleolus   a body in the nucleus involved in rRNA synthesis; it is usually associated with secondary constriction of certain chromosomes.

overdominance   a situation in which the heterozygote exhibits a more extreme manifestation of the trait under study than does either homozygote.

penetrance   the relative incidence of a trait in a given genotype.

phenocopy   an individual who has a phenotype similar to that produced by a certain mutant genotype, even though the individual may not have that genotype.

phenotype   the observable characteristics of an organism, resulting from the interplay of the genotype and the environment in which development takes place.

phenylketonuria (PKU)   a genetic disease, transmitted by simple Mendelian recessive inheritance. The gene alteration causes low activity of the enzyme phenylalanine hydroxylase, leading to a toxic accumulation of phenylalanine and of its metabolites.

plasma (blood plasma)   the yellow fluid in which red blood cells are suspended. Pure samples of plasma are obtained by adding suitable agents to prevent coagulation and then sedimenting the red cells (usually by centrifugation). See **serum.**

pleiotropy   the capacity of a gene to influence a variety of phenotypic traits.

polymorphism   the occurrence of two or more alleles for a given locus in a population, where at least two alleles appear with frequencies of more than 1 percent.

polyploid   having a simple multiple (greater than two) of the haploid number of chromosomes.

proband (propositus, or **index case)**   the affected individual who first brings the attention of the genetic researcher to a particular family.

prophase   the first phase of mitosis or meiosis, in which the chromosomes condense and become visible as distinct entities under the light microscope.

races   subdivisions of a species, recognizably different from one another.

random genetic drift   variation in gene frequency due to chance fluctuations.

random mating   a situation in which the genetic character being studied has no influence upon the choice of a mate.

recessive   an allele that causes a phenotypic effect different from that of other alleles (dominant) only when present in the homozygous state. A trait is said to be recessive if it is due to a recessive allele.

reciprocal translocation a translocation in which breaks occur in two different chromosomes and the resulting fragments are exchanged.

recombination the genetic result of **crossing-over.**

recombination fraction a measure of the frequency of crossing-over occurring between two specific loci. Recombination fractions of linked loci range from just over zero to just less than one-half. One-half is the value expected for unlinked loci.

reduction see **meiosis.**

regulatory gene a gene responsible for "switching on or off" other genes, usually through production of a repressor that regulates the activity of the other genes.

repulsion two dominant alleles at two different loci are said to be in repulsion if they are located on homologous chromosomes. Thus, the double heterozygote Ab/aB is said to be in repulsion. See **coupling.**

segregation the separation of homologous alleles at meiosis.

segregation ratio the expected (Mendelian) or observed ratio between genotypes or phenotypes in the progeny of a cross.

selection coefficient the difference between the fitness of a particular genotype and that of a "normal" genotype chosen as a standard of reference.

serum (blood serum) the yellow fluid remaining after blood is allowed to clot and the clot is removed. See **plasma.**

sex chromatin see **Barr body.**

sex chromosomes the chromosomes (X and Y) that differ in male and female karyotypes and thus can be said to be (among other things) normal genetic determinants of the sex of an individual.

sex-influenced trait an autosomal trait that appears predominantly in the members of a specific sex.

sex-limited trait a trait that is expressed *only* in members of a specific sex. This limitation is due to anatomical or physiological effects rather than to sex linkage.

sex-linked trait a trait that is determined by genes located on the sex chromosomes.

sibs (siblings, or **sibship)** offspring of the same parental combination. Brothers and sisters of a proband.

sickle-cell anemia a recessive hemolytic anemia found in homozygotes for hemoglobin S.

sickle-cell trait a condition detectable by laboratory tests but basically nonpathological, found in heterozygotes for hemoglobin S.

somatic cell hybrid a uninuclear cell that results from the fusion of two somatic cells in culture.

somatic cells cells that are not part of the germ line.

species a set of individuals who can interbreed and produce fertile progeny.

sperm (spermatozoon) the male gamete.

suppressor a mutation or allele that suppresses the phenotypic action of an allele at another locus.

Tay-Sachs disease a degenerative brain disorder of infancy due to an autosomal recessive allele in the gene controlling the enzyme hexosaminidase A. The age of onset of the disease is 4 to 6 months. Affected children show progressive mental deterioration, paralysis, deafness, blindness, and convulsions, leading to death usually between the ages of 3 and 5 years.

telophase the final phase of mitosis, when nuclear envelopes form around the newly partitioned chromosomes.

tetraploid having four times the haploid complement of chromosomes.

thalassemias a class of genetically transmitted anemias associated with reduced production (or absence) of β or α hemoglobin chains. Heterozygotes for the allele causing thalassemia have a selective advantage in malarial areas.

translocation a chromosomal abnormality in which a chromosome (or portion thereof) becomes attached to another chromosome. A balanced translocation exists when an individual (or gamete) with a translocation carries the same number of copies of each locus as exist in the normal diploid (or haploid) genome.

triplet a sequence of three nucleotides in a polynucleotide. Each triplet (except for nonsense or chain-end triplets) codes for a particular amino acid.

triploid having three times the number of chromosomes in the haploid complement.

trisomy the presence of three chromosomes of one type in an individual. The most common in humans is trisomy 21 (Down's syndrome).

Turner's syndrome a syndrome caused by monosomy for the X chromosome in the absence of a Y chromosome; the sex-chromosome complement is symbolized as XO. Affected individuals are phenotypically female, but in most cases have underdeveloped gonads. Other physical and behavioral abnormalities may be present.

X inactivation the genetic inactivation of all X chromosomes in excess of one, taking place on a random basis in each cell at an early stage in embryogenesis.

X-linked gene a gene located on the X chromosome. A trait determined by such a gene is called an X-linked trait.

Y-linked gene a gene located on the Y chromosome. A trait determined by such a gene is called a Y-linked trait.

zygote the primordial cell of a new organism, formed by the fusion of an egg and a sperm.

Index